GETTING THROUGH THE WILDERNESS:

THE FUEL CRISIS, GLOBAL WARMING, AND THE HYDROGEN
FRONTIER

BY

MARY ANN SEGAL, M.D.

authorHOUSE™

1663 LIBERTY DRIVE, SUITE 200
BLOOMINGTON, INDIANA 47403
(800) 839-8640
WWW.AUTHORHOUSE.COM

First published by AuthorHouse 12/21/05

ISBN: 1-4208-9678-4 (sc)

Library of Congress Control Number: 2005909886

Printed in the United States of America
Bloomington, Indiana

This book is printed on acid-free paper.

Cover illustration, reprinted with permission of National Geographic Society, is artist's rendition of Heronemus Windship, 12/1975

9/10/82

BEAT THE DRUM BEATS

Beat the drum beats for your survival.

Sun beats down on streets that paved

With glass, shimmer in that sun.

Sun of summer, after autumn

Nipped our buds of dreaming,

While far away the trees think

Of wrinkling up their faces in

Blush of color; fruit and gourds

Ripen into color; yellow buses

Begin to roll along the roadways

Of simple life. Children dance

With the excitement of new teachers,

New games, new accomplishments.

Our children dance, too. Dance at

The thought of name-brand jeans:

Teddy Bear, Tight-fit, Too-Too.

Tenderly I breathe the life

Abundant on the streets, watch

The spirited joy of a boy

And a band of corn-rowed little

Heads, hoping the thoughts

They think will continue to

Bless them as they reach

The Autumn of their lives,

And fruit stores well through winter.

From
JOURNEY INTO THE NEW JERUSALEM, WHICH IS NEW YORK

[Poetry by Mary Ann Segal, M.D., with Photos by Maria]
1985--Unpublished

DEDICATED TO CHRISTINE
Who listened and discussed, as I developed expertise,
And helped to keep the boat from rocking as I tried
To get ahead of the crisis

TABLE OF CONTENTS

CHAPTER ONE: INTRODUCTORY:

AN EFFORT TO AVERT CATASTROPHY

CHAPTER ONE: INTRODUCTORY:
AN EFFORT TO AVERT CATASTROPHE

I left my evangelical Episcopal church yesterday feeling calm--it's about a year since I had extensive talks with a Kerry volunteer staffer, who was passing my thoughts along to him, personally. I was discussing the fuel crisis and the alternative fuel, hydrogen. I was discussing them with many people on the Upper West Side of New York, too. People were very concerned. It came to the point that there were no more agnostics, or very few, on the Upper West Side--one man answered that question by saying, "They are all believers." I guess I mentioned God along the way, even if only to say "God bless you." I wrote to political leaders, but decided I should get on with my book. I always feel calm after church and have been wanting a month of Sundays. I found myself there, again, last June 20. It was exactly the right church for me. I stopped to play with a pretty little three year old girl just before going into the house. For the first time since 2003, I had the strong feeling that this child might live to maturity. That seemed odd, as I had just digested the impact of the prediction in the *Observer*, [in the U.K.], 7/3/05, that Matthew Simmons had made. "The price of oil may well hit $100 in six months." He is the controversial Texas oilman and banker, who was an advisor to the Bush and Cheney, at some points. Though the G8 talked about global warming and alternative fuels, it did not address the problem of *declining* oil supplies--we could all agree to the problem of *increasing oil demand, with given, or limited supplies.*. He opined that it was too little, too late to do anything about alternative fuels. We face an unprecedented crisis.

My reaction was first to worry about people staying warm this winter. About a week later, I realized that this is urgent! I was relieved it didn't happen last season, so seriously, but I did expect it to happen. However, I, too, was lulled into thinking I had time to finish my book, *Getting through the Wilderness: The Fuel Crisis, Global Warming, and the Hydrogen Frontier.* [Author House, publisher] I thought I could make a series of talks to arouse interest and action, over a period of a few years. Yet, last spring I knew better. Is it too late for my book? If this piece is published and I quickly finish the manuscript, both will become available, only around the time of the winter's high prices beginning. The problem is high prices are just going to get higher and higher as time progresses. Not like after '79 when new oil became available from the North Sea and other non-OPEC sources and prices dropped, and for many years we forgot efficiency and conservation. Are we going to panic now, after many of us have ignored the rising prices, where we could afford to do so, went on buying SUV's today [July, '05] at employee discount rates, and drove our cars through intense traffic jams to

NYC and other major cities, not using public transportation? Is the shock going to panic Wall Street?

Well its time to conserve, radically; since there has been a lot of unnecessary driving, it is not too late to benefit from conservation, SUV's and all, in our laps. But conservation has to be permanent. It is in the national interest to declare a major emergency and to stop unnecessary use of fuel. This will lower the price for necessities and help us use the limited world supply more gradually, as we prepare to make alternatives a reality. The market price should not govern investment in alternatives, especially hydrogen, from wind and solar, wind-generated electrolysis being the least expensive way to derive hydrogen, according to a U.S. DOE study, published in Feb. 2004. [*DOE Hydrogen Posture Plan,* p. A3]

I've been following the fuel crisis carefully since 12/02/01--No one else that I've encountered read the article in the *N.Y. Times Magazine, 2011,* by Niall Ferguson, an economist. He recounted predictions from Kenneth Deffeyes, oil geologist at Princeton, and from a group at the London Royal Society in October, 2001, that oil prices would be out of reach by 2010. The full weight of that prediction did not hit me till Saddam Hussein threatened to blow up his wells--we could be facing death for part of the civilization from freezing and starvation. I did learn a great deal about wind and solar hydrogen fuel, that it is utilizable in any kind of fuel burning equipment, and the fact that it could be used in any car on the road, with minor adaptations to a dual tank gas/hydrogen car. BMW has perfected dual tank hydrogen/ gas internal combustion motor cars and Boeing-Lockheed looks forward to using hydrogen in jet planes. NASA has used hydrogen since the fifties in a fighter jet. It was used in our space ships from the time of the moon shots. The *fuel cells* were in the living space, *only on board for electricity.* The power to thrust them well into outer space was made with liquid hydrogen fuel in combustion with liquid oxygen. Hydrogen is light, powerful, and safe.

I had incorporated my Op-Ed, unpublished, into my first manuscript, which is now being recast to match the speed of developments we have encountered. *World Wide Oil Crisis: A Formidable Weapon of Mass Destruction* was the title of the Op-Ed. It was submitted but not included in the *N.Y. Times,* [along with scillions of others] around the time of the war's starting in 2003.

Then, on February 24, 2004, the story on the front page of the *New York Times,* had the headline, "Forecast of Rising Oil Demand Challenges Tired Saudi Fields". It was predicted by some officials that we will "probably have a significant energy gap." Exxon-Mobil predicted that by 2010, we'll need 50% of our supplies from new fields and reservoirs. If that is so, the last, what would we need in a year, two, or three?

Well, as I agonized to try to complete a book, it was being reported that almost every country in the world is pumping out at top capacity, that no new wells had been drilled, the price not being high enough. The demand rose more than 3.4% in 2004 and had risen 2.2% in 2005. The traffic on the Fourth of July was greater and drove further than ever before, according to AAA. And the price went from the 30's to the 40's, to the 50's and recently broke sixty. My publisher said, "Mary Ann, a book that comes out in a year is going to be out-of-date." We settled for brainstorming together for awhile. The government Office of Personnel Management is giving cash incentives for carpooling, using the Metro, encouraging corporations to telecommute, and aiming in the direction of a four-day work week. These and other measures need to be used all over the nation; and others, like using the car for shopping once per week, rather than seven times, increasing availability of public transportation on an emergency basis for commuters to enter all of our major cities and leave their cars at home, and whatever else people can think of. BP had an advertisement on Saturday T.V. They asked a young woman if she would consider giving up her car.

"That would be like giving up chocolate!" she answered. " Well, it's not necessary. We have a new, low emission fuel." "Your chocolate or your life!", I say.

Conservation of highway fuel would be a great percentage contribution, to using our limited oil at a reasonable rate, at the same time cutting global warming emissions. If we did so, the price would come down so we'd have oil at a reasonable price for real necessities and for the infrastructure of hydrogen fuel and other alternatives that we deem useful with careful analysis and discussion of possibilities. We need 100% replacement of fossil fuel since it is time delimited, demand delimited. China, especially, needs to cooperate in the conservation. Their demand has increased about 40% recently, though the U.S. leads the world in massive oil consumption, 25% of the world's demand. Our natural gas prices are climbing some more this year, as its depletion parallels that of oil. That is less easy to conserve but houses can be better insulated, evaluated by professionals. Businesses, too, have to improve their efficiency standards, to survive the climb in natural gas and electricity costs. It is urgent for us to have clean, powerful fuel that can replace petroleum products, and natural gas. Coal is cheap and plentiful for electricity, but it is extremely polluting. A potent energy fuel is needed in spaceships, jet planes, ships, factories, electricity plants, and cars. The only fuel that could have this energy potential is hydrogen—in smaller amounts, it can and has helped in home heating, and in household appliances. It can even be used in a Coleman stove, on a camping trip. My precious metal jeweler friend says that his colleagues plug into the wall, electrolyze water and use a hydrogen "water torch" for melting metals, as in soldering.

4

To get to the place that we can put hydrogen into use, we must keep our fuel supply and prices adequate for necessities. China must slow down--we can help by not needing so much from Wal-Mart. We don't need Fijii bottled water, either, or a number of things that are freighted too far with too much fuel. This should involve the rich and the poor alike exercising conservation. We are at war with our poor planning, with not having made the transition in earlier decades to renewable, clean hydrogen. Nothing could prevent outstripping of the oil supplies, living with an ever expanding population, growing demand, sudden world-wide industrialization—because finally the oil would peak at the half way mark of the amount in the earth, the last half being increasingly difficult and expensive to obtain, less and less after a plateau, perhaps, that we can cope with if we have the sense to *plan abruptly and now,* making it possible for the luxuries to be foregone as un-American and insane, while we count what oil we save for necessities, as carefully as we are forced to count our money if in the lower-income brackets, and as carefully as we count our profits in the corporate world. If we lose our grip on the care with which we use our oil, we shall lose our grip on having a civilization as we now know it.

.   .   .

Mary Ann T. Segal, M.D.
July 20, 2005

President George W. Bush
The White House
Washington, D.C.
Fax: (202) 456-2461

Dear President Bush:

I am most distressed since last weekend, when the dawn broke upon me, that the enclosed article from the *Observer* regarding oil is an indication of what I was thinking a year ago--the panic point for fuel consumption would be sooner rather than later. My book has been on hold as I gathered more information. Now it is on hold, because it can come out in maybe four months, but that might be too late to warn people in time to conserve fuel. Before I return to writing, I am contacting you and other people in positions of political leadership, who are capable of addressing the nation and the states and the regions.

Obviously, what you announce now will be far more efficient than what I just begin to communicate in several months. The policies that are being practiced by the OPM and corporations in Washington, D.C. are appropriate for the whole nation. Car-pooling, use of public transportation, telecommuting, and shooting for a four day work week. I heard about this upon making a call to SBA. Cash incentives? How about tax deductions? Or just plain, We have to avert an emergency and it is necessary to cut highway fuel as much as possible. Though it is the vacation season, what we conserve now will offset very high winter prices, if we work at it for the next five or six months? Every effort will be made to increase the availability for commuters into our major cities of public transportation--those using it can deduct half their expenses from their taxes. Try to substitute unnecessary car trips with bus or train trips. Organize shopping trips so that you can make one instead of several.--Long term fuel is limited, but can last much longer if we conserve. This conservation will cut much emission, help us know that we can control this problem to a large degree, and give us time to get alternatives into place.

The discussion should come up now, before the recess. I can't describe the anguish I've been feeling over this problem. Thankful we got through last year without calamity, I got lulled into thinking we had enough time for me to publish and speak at many places to organize interest and action. Now, I highly doubt that. My article, enclosed, for *Christianity Today* cannot even come out until November, if they take it, for "Speaking Out." I'll send that to them. The book could come out by November, or so, too. But completing the book is not as important as contacting you, and others in public authority. It can be handled at the state and local level too. Please let my suggestions help. This is dreadfully alarming, but I gravely doubt that it is alarmist.

Sincerely yours,

Mary Ann T. Segal, M.D.

Enclosed: *Observer* Article (2pp.)
          Article for *Christianity Today* (2pp.)
          Letter to Assemblyman Keith Wright (3pp.)

CC: Dr. Howard Dean, Chairman, DNC

     .   .   .

Hydrogen can be produced at a scale to replace all fossil and nuclear fuel, with wind/solar, equipment. I justify that in my book, drawing on what I learned, particularly from Harry Braun, analyst of the years of research, many decades of research, since the beginning of the last century, on solar and wind-generated electrolysis of water. His book, *The Phoenix Project: Shifting from Oil to Hydrogen* (2000 ed., SPI Publications) is extremely informative, as was the 18 months of discussion he shared with me, perhaps twice a week. Like any website, [phoenixproject.net] holds some of his knowledge, but I advise strongly that one read his book. I will write of other potential pieces in the variety of possibilities for alternatives. I shall include other information about hydrogen energy, gleaned from the *International Journal of Hydrogen Energy,* and other sources. There is a huge amount of research done, and being done on this great fuel, non-polluting, if not derived from fossil fuel, but from water. I hope to include enough about it to give the reader a taste of freedom in learning about the potential for vast amounts of clean, renewable energy. We must start with it before it's too late, however, leaving some fossil fuel for the products we make from it like medicines and asphalt. Naturally, there is resistance from the oil and coal people, from politicians, but one chief reason is that we, the public, are not aware of it sufficiently to demand it.

An army of Christian soldiers, and of other concerned groups and networks, can help themselves and others to conserve, and urge the state, local, and federal governments to assist with this. This is the most serious crisis our civilization has known. It can be controlled, eased, and we can move on to the hydrogen frontier, enriched with solar heated homes and ocean current generated electricity. We should be looking for a 100% replacement of carbon-emitting processes, while we look. Conferences to sort out the best contributions should be held, with cross discussion where we can. And as the Institute for Analysis of Global Security has begun, we should have many cooperative conferences with China, in this emergency.

CHAPTER TWO:

WAYS TO CONSERVE FUEL

The objectives for conserving fuel are to control the climbing demand that has met its match in the ability of the world's oil companies to supply such demand (from the U.S., 25% of the world's demand, fast being very high from China, then from India.)

This puts us in a position that if an unforeseen political or supply problem develops, we are vulnerable to gaps in supply for some necessities. Unprepared, we could become chaotic, and panicked and potentially unravel the economic and food/heat/mobility structures we depend upon.

A second objective is that the price would go down, and be reasonable for what is a fuel necessity. It can stay down, if we develop new habits, in its use. It would be at a much more affordable level, if we have a consensus to build the infrastructure for wind/solar hydrogen production. At the current demand and price level, fuel for such manufacture would be a further burden on supply and would have to be done at a cost that we needn't make so high as it is at present.

A third objective is that not using as much road fuel, the biggest source of fuel use, we will have made a significant dent in carbon dioxide and other polluting emissions. Our new habits are not *that hard to develop.* We shall at least be able to know, that for the time being our situation would be under control, while we think of new ways to improve the picture.

1. Washington's Office of Professional Management has instituted the following, applicable to some major cities: Cash incentives to use carpooling or the Metro public transportation system. I believe that a city with a good public transportation system should be utilized to the fullest extent--it saves more fuel than carpooling. Services would have to be expanded for the increased volume of passengers, and should be done so with the greatest efficiency. The Mass Transit Bill, passed in July, 2005, before the recess, provided public transportation from mass transit stops, to the job site, for federal employees in the D.C. area. A member of the American Public Transportation Association informed me that transportation from major public transportation stops to the job site is being provided in some other cities, as well, and is not unique to Washington, D.C.'s area.

The OPM and other companies in D.C. are encouraging "telecommuting" for people who work on company computers and telephones to work at home. agency phones could be call forwarded to the employees home, perhaps, without revealing the privacy of the number there.

Some four day work weeks are being set up, by having workers be at work nine hours a day, for example. [This is a verbal summary of one employee's grasp of the overall program, not official from the OPM website.]

I'll also note that a Gateway representative stated that the N.Y. office was considering having their representatives work from home. He did not know the outcome, but while we were talking, realized that many people from his city go to the same worksite, at the same shift. He saw the possibility of carpooling as we talked.—On further consideration, perhaps there are enough of them going to the same worksite, at the same shift, to hire a van or bus, through the public transportation funds, or as a group, with tax deductions for their expenses.

2. Public transportation exists for the elderly in some rural areas, though not all. The same system could be used for many people besides the elderly to get to the cities where they shop for food and other needs. I visualize this in western Massachusetts and central Massachusetts, for the smaller towns. The distances are not vast, as they might be in the rural areas of some states. Each geographical area needs to assess their own regional needs. John Neff of the American Public Transportation Association has kindly supplied resource information about how local people, leadership committees perhaps, could investigate if there is current funding for needed transportation additions. This association of persons in the transportation sector have been improving the sector for years and have a goal of an annual 3.5% increase, in perpetuity.

3. Why did the express lanes in CA that I saw on T.V. have neither carpoolers nor people paying $11.00 on them? They seemed empty. I'm not familiar with the transportation mores in California, except to know they are in gridlock. Can multiple buses pick people up and speed up the flow of traffic? A colleague reported to me that when the pain started [addition, 10/4/05], a man entering New York City, on a highway with a fast lane, had people gather that he could pick up, thereby carpooling without difficulty finding people and able to use the fast lane. It would be a help to have people gather at sites on roads, where willing drivers could fill up their cars, with or without a fast lane.

4. New York City has an excellent public transportation network. Most of us use it, though many cars and SUV's line our streets, too. Hybrid buses double the mileage and have been enlisted to an increased degree. Yet we still have many commuters into this vast metropolis of employment coming to work in cars, tying up the roads into the city and exiting, tying up the toll entrances, and the rush time traffic within the city itself. If there were enough extra train cars or scheduled trains, reinforced by more readily added buses, from Connecticut, Westchester, Long Island and New Jersey,

these daily frustrating traffic jams would dissipate, making the commute less frustrating and time consuming, and conserving fuel at the same time. Once it were running smoothly, I don't think anyone would prefer the old ways. It's not like getting in your car, said a man in bus transportation public relations. But, he added, that it was wonderful to be in Germany where the bus system was well planned and there was never a traffic snarl.

Though heating fuel shows no increase that is not weather related and is not so easy to conserve as highway, or airline fuel, or private jet fuel, there could be considerable savings in thirty buildings to a block, if for eight hours a day, the people who don't work, congregate and socialize, with spontaneous or directed activities in churches on the corner that could serve as community centers. The people who stay at home are unemployed, disabled, elderly, or other. Many can't wait for pleasant weather to congregate socially on the streets. So the winter months may be ones of loneliness and isolation. People who could not leave the building for exceptional reasons, could heat their favorite room with a space heater, the cost coming from the general savings on heating fuel. Similar blocks exist throughout the five boroughs--drawing on heat for millions. Legal provisions could be made so that there is an equitable benefit for the tenants who cooperate with landlords in this way. Care must be taken that no one is forced out by disreputable landlords, who chronically don't heat. Cash incentives may well be available for the participants, most of whom are poor. Landlords could receive the funds necessary to adjust their heating systems to trigger appropriately for the circumstances. This may sound like sacrifice, but I think it could be a lot of fun for those involved. Cumulatively, it should save a lot of fuel oil. I polled at least 15 people on my block in the spring of 2004, regarding this measure. Many were interested in cooperating with this plan, in a crisis. Last winter was not a crisis, but this winter may be.

This takes tremendous organization--perhaps it could be done by block-association members in cooperation with leaders that people accept as leaders, from the local churches that would be housing for the day programs. It may be a way to avoid passing on fuel costs to tenants, as was done in the past.

[There is the possibility that this might not conserve sufficient fuel, for eight hours per day, as it is recommended that someone leaving the house to go to work, or in this case the community center, should not lower the heat more than 10 degrees below the comfort zone for the evening. To lower it more than 10 degrees is said to require more energy to bring the temperature back to what is comfortable.

The alternative to the five day a week, eight hour community centers, in the neighborhood, would be what I am recommending for the rest of the country, envisioning the simplicity to be greater in smaller communities. That alternative would be to form five day/24 hour shelters, in churches, schools, and community centers. This goal may be a necessity in all the colder regions of the country, including the boroughs of N.Y. With high gasoline prices, higher electric costs [which are coming, according to Con Ed, though they can't quote a percentage increase yet], and higher gas, electric, and oil heating bills, many people will have heat-and-eat problems. Not only would our bills go down, and we could have some money to buy what is possible, keeping the larger economy going, we would also create a reserve of heating fuel and methane for electricity.]

5. Summer [or any season's] trips by train, subsidized by the government already, could perhaps have lower fares. One need not go forever on the train to get to an entertaining destination. The kids wouldn't be saying, "Are we almost there yet, Daddy?" as often, if they could go to the food car for distraction and not get stuck in traffic. The train ride from NYC to Boston was absolutely beautiful--and too short for me last summer. The cross country lines are greatly underutilized. Buses wouldn't suit family comfort as much as trains for a longer trip, but *are* more affordable. Trains have leg room, toilets, food, and the chance to stretch your legs with a walk. AMTRAK charges ½ fare for every two children accompanied by one adult. I priced a trip for a family of four from Chicago to Disney World. I am quite certain that it was less expensive than air fares to that destination. Family bedrooms were said to be very comfortable, both for sitting and for sleeping. Disney World is worth the time in the train for a well motivated group—jet fuel is very voluminous in use. Railroads have little friction, with their steel wheels on steel tracks, so energy use is low compared with driving, and certainly with flying.

6. Some pre-owned luxury cars have good gas mileage, good handling, and an expected lifespan that will continue, with prices to match new GM's and Ford's. There is a long waiting list for the Toyota Prius and the other hybrids. Some people in New York City have said they might give up their cars. Those in good condition, with higher than SUV mileage, may be needed by someone in another area, where getting to work is not possible by public transportation, and finding an affordable second hand car that's dependable is perhaps a chronic problem for those with a low income. Perhaps someone wanting to put aside their SUV with low mileage would also be in the market for a car that was deemed no longer necessary, by people who enjoy driving, but could just as well not spend that money where public transportation is available. There are the insurance, the tickets, the repairs, the search for parking, the digging out in winter, and the gasoline to be considered. An emergency trip, on a day when health or strength made public transportation too much of an ordeal, would be affordable with a parked-if-idle car service, called upon request. Cruising empty taxis are a burden on fuel use. Logistics could be worked out to minimize the number of empty taxis in circulation, who

are overpopulating some areas. The habit of using public transportation, instead of taxis, should be developed. If this is a hardship, taxis could be parked at standing areas to walk to, or available upon request by phone arrangements. At the train stations and bus stations, taxis and red caps could be plentiful for people using railroads and interstate bus lines, who carry a lot of baggage from a long trip *without their cars.*

. . .

I have been discussing all that I can think of to save fuel with the cars we have at present, those which might be available to buy. I have been discussing a concerted effort to use public transportation to the extent that it is available, and I advise a great outcry for a more logistically complete public transportation system. There is no news of a great deal of new telecommuting. That is another great way to save fuel. There seems to be no problem to outsource and employ people in India with some of the companies I do business with. Why should telecommuting be so sluggish? Companies talk of security issues, but I believe there are other solutions to that problem. What is so private about being constantly told to give your social security number over the phone.? Companies will not give me my account number over the phone. What's different about that risk? It's an indication that they don't consider the phones secure. There are not many jobs that can cut down to a four day work week that I can think of. But Washington, D.C. is considering it for federal employees.

Let me mention now that obviously hybrids are good for the near future, but there is a new piece of important news. A hybrid can be equipped with a battery, new in development and strength, that can be charged with the electricity at home, and permit travel on the strength of the battery alone, to ride about town shopping, or going to church, or school, or a movie, without ever using the fuel in the hybrid at all. Prius+ is the name of the battery added to the hybrid Prius. They will be available in 2006, the last I heard. Of course, if your job is fairly close to home, you may be able to get to work and back, without ever burning any fuel. The car could be charged up on off-peak hours. The long range problem with this technology is that if the demand is great enough for that kind of battery charging, there will be no off-peak in the heat of summer. Furthermore, the cost of electricity is rising with fall of supply of domestic and Canadian natural gas. And we don't want to burn more coal. The other drawback to this technology is that not everyone can afford it. But acutely, if may help with our fuel emergency.

It should be clearly understood that liquid hydrogen tanks and fuel injection lines can be added to any vehicle on the road, at very low cost, easily subsidized, just as the Prius+ is subsidized by the government. That vehicle, car, truck, bus, and hybrid car can have dual tanks, one for gasoline and one for liquid hydrogen, to burn in an internal combustion motor. They would be modeled after the BMW 745h model, to be marketed by 2008. This carefully engineered internal combustion motor that utilizes either gasoline or hydrogen, with the flip of a switch, has been researched since the 1970's. <u>It is not a fuel cell car. It does not cost a million dollars. It would cost about 25% more than a mass produced BMW, or 5% more if mass-produced, it was estimated in the past. I don't have current figures or projections. I do know that BMW's goal is to partnership with companies in every country, to make this new frontier possible.</u> It is the goal of BMW and others to make liquid hydrogen available as the fuel of the future. If every car on the road used hydrogen, there would be no further pollution from cars. It is not necessary to get all snarled up with fuel cell cars that will probably never be affordable by the masses. [Based on reading, *ibid.,* and discussions with Harry Braun.]

Yet that is about all people have heard about in connection with hydrogen. You hear hydrogen, you hear fuel cells. The fuel cell batteries have to be replaced every year. For the standing fuel cell electricity in a single home, the cost is $100,000 dollars, and the battery has to be replaced in a year or two, working toward three. We complain that hydrogen infrastructure will cost a lot, yet the governor of New York leased a Honda Fuel Cell, in the name of alternative fuel, probably not aware of BMW's technology. A recent Automobile section of the *New York Times* explained a great deal about the Honda Fuel Cell car, stating it would cost from one million to two million. At the end of the article, there was an inconspicuous paragraph about BMW's dual tank internal combustion motor. Both use hydrogen and emit water. The fuel cell doesn't burn the hydrogen; it draws electrons from it to make electricity. In this way, there is no small amount of nitrous oxide resulting from the heated air, that contains nitrogen. The internal combustion engine does heat the air. If properly controlled, there is no danger of nitrous oxide emission. Under other conditions, BMW's press release paper (2004),explains that a nitrous oxide scrubber would be built into the exhaust. Rest assured, scrubbers can be built into all liquid hydrogen burning equipment. In another place in the BMW Group Media Information (9/2005), it was explained that the window in which nitrous oxide is formed, temperature-wise, can be avoided by toying with the rich and lean air mix aspects to hydrogen fuel.

Liquid hydrogen burns effectively in any form of fuel burning equipment, from the Space Program, to Boeing-Lockheed engineered jet planes, to ships, to diesel trains, to electricity plants, to factories, steel mills, trucks, cars, household appliances, Coleman stoves, and jewelers' torches. Hydrogen is light, powerful, and safe. NASA stated that it is the safest of fuels in a highway accident. I will expand upon the safety issues of hydrogen in detail, as we are all left with the erroneous memory that the Hindenburg exploded and hear the terrible misnomer, the Hydrogen Bomb. [The explanation about the Hindenburg is in Harry's book, and was also referred to in the *Journal of the International Hydrogen Association,* the latter article explaining that someone on NASA's staff did the investigation of the matter. See Chapter Three.]

I think that it is very safe to say that there is no other alternative fuel to keep our jet planes flying, none other that is sufficiently powerful. Remember, NASA used hydrogen fuel in a jet fighter in the fifties. From there, it being so light and powerful, it was used in the Apollo moon shots, and ever since, getting a spaceship to Jupiter. As there is no air in space, beyond our atmosphere, liquid oxygen is on board to combine in the fire with liquid hydrogen. The Germans used hydrogen in submarines in the thirties and forties. It was generated from water by offshore windmill electrolysis.

I shall discuss the Prius + in the next chapter. Along with it, I will discuss the perfect substitute for ethanols. There is a piece of equipment to add to a car, patented in Canada, that does on-board electrolysis of water, sufficiently to be able to claim, under experimental patent that it gets 21% reduction in the gasoline use. It seems too good to be true. But for $197, one can equip any car with this equipment, and I believe beat the amount saved by mixing ethanol with gasoline. Yet all it requires is a one gallon tank of water, holding an electrolyte, lye orbaking soda for example, and the proper electrical equipment to allow the running car to be able to allow sufficient electricity to pass through the electrolysis tank, to release an oxygen rich, hydrogen mixture to mix with the gasoline, increasing the mileage by at least 21%. It is adjusted so as not to reduce the horse power. I was very surprised to find this invention on the web. Surprised and delighted. It's a help, and it's so simple. No massive cornfields, fuel burning tractors, fuel burning irrigation, government subsidies and loss of much needed cropland. Just a water tank with a simple electrolyte, drawing electricity from the running car. A 21% reduction in automobile usage of gasoline is significant. It is adaptable to all kinds of cars and trucks, SUV's, compact cars, even a hybrid. With the latter there was a smaller amount of gain, yet there was some. The selling cost of the equipment is currently only $197, at [savefuel. ca]. I have discussed this equipment on three occasions with the engineer, so interested in hydrogen, who worked on prototypes till this was perfected several years ago.

Obviously, increasing fuel efficiency standards for our American cars, is an urgent need, but the slated first year is 2007. I read an alarming report from the Sierra Club about the proposals from the Administration—There is only but about a 2MPG increase required, and vehicles of larger sizes are in a grade of sizes, exempted from better mileage requirements. No change while we are here in the midst of a fuel price pinnacle—and as the Sierra Club writer said, farmers need trucks, without doubt. What option do they have, except to buy what is manufactured, poor technology for fuel use though they have. Families of any size more than five have needed the only large vehicles on the market, SUV's. I asked someone the other day, when was the last station wagon advertised? He was young enough not to have known what a station wagon is. I met a man in Massachusetts, who had a house in Rome, Italy. There he had an Opel station wagon with computerized fuel injection that achieved 80 MPG. Now, someone else said, they aren't powerhouses. But they do achieve the job of transporting a family. Our 55 mph speed limits do not require tremendous power.

If carpooling is hard to arrange, people could group along the road, at agreed upon stops, or in a local parking area where they leave their cars, and make themselves known as people who wish to carpool. That way an empty car can stop and pick them up. One fellow on a New York City entry road does this. The rules for him are, nobody discussed paying, no one talks, and the radio station stays at the driver's choice of stations. He can get into the fast lane, but if there is no fast lane, a lot of fuel has been saved. If buses are possible along this road, and people used them, there would be less traffic-jamming since there would be fewer vehicles. The buses give more conservation when they are well subscribed.

The Mass Transit provisions from the bill passed before recess, on July 29, 2005 are more generous than past bills, but do not provide for the fuel emergency, as I conceive of the problem. APTA, the American Public Transportation Association, however reported that the bill was a historic moment, and helps towards their goal of annual increase in public transportation availability of 3.5%, in perpetuity. I asked John Neff, of that association, to tell me how APTA might advise regarding citizens requests for new public transportation or increases in all ready available public transportation. He sent the letter that gave numerous websites for different circumstances, to contact and see if provisions are available already to put extra services that you may want in your area. Unfortunately, I could not successfully scan it for publication. I would advise people who think of improvements or additions to public transportation in their area to contact the APTA. Contact information is available at [apta.com.] Your local transportation authorities may also be of help. Another resource would be local political leaders.

Community committees or leaders of groups concerned can lead the inquiry— I also know that private buses can be rented by individuals who are licensed to drive a bus or transportation van.

CHAPTER THREE:

LETTER TO AN ACQUAINTANCE

LETTER TO AN ACQUAINTANCE

Mary Ann Segal, M.D.
New York, New York

September 10, 2005

Dean Barry:

I wrote you a long letter, starting on August 29. I had been preoccupied about our conversation. And puzzled. I enjoyed writing you and got very caught up, for a while deciding it should be my next chapter, but subsequently deciding that it was much too theological and heavy for the general reader. Perhaps I'll send that to you, but I'm thinking of this as perhaps my third chapter, so am reframing it so that it fits everyone's experience a bit more readily.

This manuscript is new and urgent. What I had prepared for Beacon Press is somewhat like a dinosaur. The editor with whom I was talking and I agreed that a manuscript which came out in a year from being finished, would be out of date. I had read all the books, or most, that have been dealing with the predicted end of oil, along with many, many articles. I read the *Boiling Point,* by Ross Gelbspan which enumerates the many signs of problems progressing from global warming that have already occurred, I read much of Jim Wallis' *God's Politics: etc.* I thought I had read it all. Christopher Vyce, assistant to the director at Beacon said, "Mary Ann, there's a book I want you to read." It was *The Long Emergency,* by James Kunstler. "It's very depressing," Christopher said. Well, I got the book, and read it over a weekend. I was filled with grief, not depression. It was the author's attempt to describe a shattered America, living through the pestilence of complications from global warming-without fuel-and finding pockets of survival, built around the remnants of Adirondack Park towns, and similar places , going back to local farming and cottage industry where the children would be farming more than schooling. Walking would be the main means of transportation. Implied was that many had not survived. I handled the pestilence part like a doctor on constant call, if I lived through it, not feeling defeated, but rather, needed. After all, that's how Doctors Without Borders live. The local farming, the kids having short childhoods, contracted by their chores, struck me like my own childhood, except there was no need not to have school, and plan to go to college, for me. There is plenty of time to work after school. There were myriads unaccounted for after he gets some of us settled in these great old towns.

Then I stopped suffering when he got to laughing at the feasibility of hydrogen fuel. He just didn't know anything about hydrogen as I have studied it. So the crisis was over for me. I read on however, and saw him trace the use of energy through the ages, and the development of a greed and manipulation economy. That made me very anxious, as that is a reality which complicates our lives. I spent three days in grief, for if we didn't find an energy economy to suit, his pessimism would be appropriate. I left Christopher a voice mail about my reactions to that book. He called me several days later and told me about the statement of Matthew Simmons's in the *Observer,* on July 3, 2005.

I had already decided to write an expedited book, self-published, and was doing more research. Then suddenly, my thoughts in an email to NASA, with questions about hydrogen, were interrupted by some loud music. I got irritated, lost my three at a time thoughts, and had to stop. Perusing what else I could do, the urgency suddenly hit me. I had to try to tell people what we could do to slow down our demand for oil, so we'd make it through the winter---not that my late suggestions would help too much for this winter, but we have to set ourselves on a sane course, or it *will* be the Long Emergency. Some of what I have to say may help us this winter. Certainly, we should not despair, for if we understand why this is happening, and begin to apply strong conservation measures, perhaps next winter can be much easier.

This winter will probably require lots of ingenuity—helping one another travel, impromptu vans on rural roads, emergency requests for public transportation, and maybe shelters with shared heating. I know that Americans are practical and filled with ingenuity in emergencies. There are many men in target areas who know how to show people how to turn off their water and drain the pipes with open faucets, so the pipes cannot freeze. Shelters in the school auditoriums for people who can't afford to heat their homes can be set up. Church meeting rooms, community centers, and VFW halls could also be used. People would go to work, could go out for recreation, at their choice. There is no need to run the shelters in an institutional fashion. To be in company, rather than helplessly cold and desperate at home, would be a lot more supportive and morale could be built around the cooperation involved. Getting this started could perhaps be done by advertising a public meeting through the city or county or town official who advocates for the needs of people. Start discussing it among friends, neighbors, church congregations, etc. as soon as possible.

<u>While plans for shelters are not prepared, individuals can have a lot more control of their circumstances if they house-pool with friends, neighbors, or relatives, sharing the funds for heating and electricity in one house.</u>

. . .

Today's *N.Y.Times* had an article about the pessimists and the optimists with regard to the future of oil. I had no quick answer for the way the optimists phrased themselves. I know from all that I've read that I don't trust them. They are right that the price, if high enough, prompts conservation, and so does its magic in time on the supply-and-demand side. The author, unnamed, said he couldn't draw a conclusion between the pessimists and the optimists. He concludes by saying the price, and the supply-and-demand side, did it's magic before. "And it will again. Until it doesn't." –Shall we wait, I say, optimistically, to make a major change in our fuel supply. [I shall return to this issue in depth, in the next chapter.] The optimist's end-time is not that far off either. It's best to get the cost of oil down and begin the infrastructure for wind/ solar hydrogen. Rapidly, as the hydrogen can be used in many applications. Just think of planes. Without the high octane jet fuel from oil, there is no other fuel powerful enough to run a jet plane. NASA made bombers in the 50's with hydrogen. And according to the *DOE Hydrogen Posture Plan 2/2004 A3,* wind turbine generation of electricity for electrolysis of water is the least expensive way to get hydrogen. Boeing-Lockheed has designs for hydrogen fueled planes, with a take-off weight that is 40% lighter. Had there been hydrogen in the jets that hit the World Trade Center, the building would never have had melted girders, the fire would have been much smaller, the smoke inhalation insignificant, for the fire of hydrogen makes water vapor. Hydrogen, the lightest element, quickly dissipates upward into the air. Hydrogen is very safe. Okay, I'll say it now:

The Hindenburg disaster was set off by lightning striking the aluminum paint, which turned out to be a jet-propellant. The hydrogen was set afire, but it did not explode. The people panicked and some jumped. Of those who didn't jump, only two died; they, burned by zeppelin's diesel fuel from its motor. That zeppelin was longer than a 747, so it was vast in size and the fire was large, but not explosive. The ship sunk slowly to the ground.

We have a lot of coal out west. It is sub bituminous coal, less energy providing and less easy to burn than eastern bituminous coal. There are lots of problems transporting it as well. I think it is very unfortunate that the energy industry is off to the races, to develop that coal into gasified coal, that being some ten to twenty years off, according to Jeremy Main, in "Old King Coal is Back", *Fortune,* [February 21, 2005.] Once the plants are widespread, carbon sequestration will be of billions of tons. None of this is economical yet on a vast scale, nor is it known how to sequester that much carbon dioxide. This is an issue that intuitively disturbs me, carbon sequestration. It seems an obscene solution, that could have many repercussions. Let me say for now that the coal would last longer than OPEC oil, but only 200-250 years, perhaps. Out of oil, out of natural gas, and out of coal, what would we do then. Write off human history?

Hydrogen, from water, is forever renewable, and in mass production by wind electrolysis, it would be cheaper than gasoline, by simply factoring in some of the subsidies for oil; and if we factored in all the health costs, military costs, and corroded buildings, bridges, ancient monuments, and forest death from acid rain, hydrogen would be insignificantly costly. Now, we get hydrogen from fossil fuel, the cost higher, and we are happy to pay the price. We don't think clearly about money when assessing costs. The Governor of New York leases a Honda Fuel Cell. He may not know about BMW's internal combustion hydrogen/gasoline cars and buses. Not many people do. But they are just as clean, and when the heat is controlled, there is no nitrous oxide emitted. If the heat of the motor makes some nitrous oxide, the exhaust system has been designed to scrub the nitrous oxide.

.   .   .

I want to shift away from the anxieties of economic ploys, from the memory of the Lord Browne of Madingley announcing to his BP colleagues, that yes, it is time for a clean fuel, one that won't cause global warming, hydrogen, made from fossil fuels. A British friend referred me to the website. He, like you, she said is having trouble convincing people to use hydrogen. In fact, I'm not having trouble convincing most people. But I have a bunch of friends who take me by surprise, like she did, telling me of a hydrogen car! [fuel cell], or in her case of a clean fuel, with a sleight of hand added, that she wouldn't understand. This energy field is in such a state of confusion! I now generally understand the confusion, but how could anyone who has not set about studying it, comprehensively, alerted first on December 2, 2001. Not to say others have not been studying it much longer. But for the man in the street?

.   .   .

Let's be people, for the moment, thankful the worst is somewhat over in New Orleans, that slow, painful recovery and relocation will take place, that we've been in shock with the unbelievable losses of life and families, and homes, and lovely New Orleans, where we all wish we'd had the chance to visit, if we already haven't. The Birth of the Blues....

I started a book on the fuel crisis—but four or five weeks into the book, we had the development of a fuel crisis, upon a fuel crisis. People of less means were already starting to use credit cards to get to work, others having trouble driving for chemotherapy on a forty mile trip every week, some of my neighbors with cars in New York City talked of giving them up, some were very angry and planned to boycott gasoline on Thursdays to get back at the false prices, thinking there is a conspiracy to raise them when there is plenty of oil...and I slowly thinking, yes, were it not that we needed a price pain to tell us of shortage in face of the demand, it could be considered price gouging to use the market supply-demand rule for a necessity. If the prices were controlled, we'd merrily have gone off on Labor Day at record mileage, predicted by the AAA, as was the case on the Fourth of July. And sadly, by Labor Day, many stayed home—shocked by what had happened.

It did warm my heart to hear of the guys getting vans organized to help people who had no gasoline, during the Katrina crisis. That's what I hope they will continue to do in less pressured times this winter. Hopefully less pressured times. That is still a matter of being a seer. Some major cable network announced that oil would be 70% more expensive this winter than last winter. That's been the trend. And getting the house properly insulated on a broken down budget is not so easy up in Adirondack Park. My nephew lives up there, as does James Kunstler. Jobs are not easy to find. Fortunately, between him and his wife, they managed last year. It's going to be a hard winter. There's a way to keep the pipes from freezing and huddling together in shelters, if the fuel is too expensive to heat adequately. All the fuel is expensive: gas, natural gas, electric, oil, and wood. Many, I suppose, will use wood, but that's not a good long range plan. Under average circumstances, it doesn't heat the house too well either.

The *N.Y.Times* also had an article today entitled "The New Prize: Alternative Fuels." Those options listed are going to be discussed as I continue my book. People are bursting with thirst for new options. The itemization included E85 [ethanol 85%/gasoline 15%] It's hard to locate and costs only 40 to 50 cents less. As a policy, broadly used, it's a definite misfit. It is most obviously a crop grown on food land and at that rate would take the whole cropland of the U.S. to provide for everyone, according to the *Times*. Lester Brown, the ecologist, would estimate more land than that. It isn't as potent as gasoline, so perhaps nothing was saved, money wise. It takes tons of fossil fuel to grow and irrigate it and process it, so it is very polluting and energy negative. I have been communicating with Dr. David Pimentel from Cornell University, who has written an article about the ethanols and their negative energy, all requiring more fossil fuel to prepare them and thereby giving off pollutants, and all coming out with fewer kcalories than went into its production.

The second item was cars running on natural gas—it's a pity that our natural gas, domestically, is declining so rapidly and that electricity shot up this summer, as will heating with natural gas or electricity this winter.

Fuel cell cars were the third item. That was mentioned theoretically, and not as though in use. The cost hurdle is too high, the pumping stations not built, and the hydrogen supply a problem thus far.

The older electric cars that have limited range, long charging times and the environmental and infrastructure problems of supplying the electricity.

Biodiesel use ranging from McDonalds used cooking oil, to soybean oil and vegetable oils.

Plug-in hybrids have been gaining a lot of attention. [Calcars.org] is a strong promoter of this solution. The Institute for Analysis of Global Security has also praised plug-in hybrids as a solution. I think the strongest momentum has been for this solution, where a battery charged at home, and at night, is technically superior to older batteries, and can carry the car for a trip to shop in town, or to and from a workplace that is not too far from home, without ever touching the gasoline.

In the acuteness of the fuel crisis, they seem very appealing. Some are in operation, but supplying customers will take some time. They are expected to reach the market by 2006. Yet they cost $12,000 more than the basic hybrid. The large problem is, that if used *en masse*, they would use a lot of electricity not currently in use, pulling down the methane supply, to be replaced as things are, by more coal plants.

I was going to sound an apology for making such a big fuss about giving up cars in favor of public transportation increases. After going down this list, with what I already know in detail, it sounds like a very important goal, even if I sound like the Witch of Windsor. I realize that not everyone has access to public transportation, or they are traveling salesmen and public transportation wouldn't work. Or would it? I guess it depends on where you are going and what you are selling. Cars are essential for many people. But organized van systems like the ones that cropped up spontaneously during Katrina, might be a way of sharing the fuel of a larger vehicle, and leaving one's own at home. I can just imagine how my sister would react to that—no way. She has a business. She couldn't really do it. And the men! It wouldn't be like giving up chocolate, it would be like giving up...think of a good metaphor. They think of mopeds, instead, and don't think of the weather like I would. Motorcycles use less gas, too. But what if you are a farmer and really need to drive the truck that gets 8 miles to the gallon. You are carrying grain in the back for the cattle. I saw a motor scooter on the road yesterday. The man spends three dollars a week on gas and loves it.

Well, forgive me if I sound like the Witch of Windsor. Think of public transportation, even if new logistics have to be worked out, or additional buses and trains have to be added. It cuts fuel use 50%. Probably more, if fully utilized. We subsidize ethanol, we subsidize fossil fuel, let's ask for subsidies for public transportation that we need to be increased. Fares held reasonable. Let's make every section of the country as accessible to general transportation as is possible. Could school buses carry shoppers to town, on the hours that they are idle?

Regarding public transportation, for everyone who *can* use it, more fuel is freed up for the neediest drivers. We are working, I hope, in the direction of making ourselves able to buy oil and gas for necessities at a price we can afford. Everyone's necessities. I'm sitting here at a computer, for weeks on end now. I have no idea of who is living where, who is reading this book. I just know I love you as a fellow American, or as a member of one world, and I know that we've been demanding more fuel than the Saudi Arabians may be comfortable with. Their wells could collapse, at least Ghawar, the one which is pumping out 30% water, if they pushed them too hard.

The China problem? Increasing their demand 40% in one year. I don't know enough about their life conditions, their need to "grow their economy" so fast. I strongly suspect that our both investing in the infrastructure for water electrolysis to get really clean and renewable hydrogen to replace coal that kills with pollution and oil and methane that are declining, could create an economy at home, thus immediately increasing Gross Domestic Product and creating many jobs. We have nothing to invest in, in the U.S., besides the oil tragedy and the housing bubble. Everything else is just about accounted for, according to Wall Street Week, about one year ago. The money was going into currency speculation. The available amount of money to invest was quoted as six hundred billion dollars, on Wall Street.

Barry, this was supposed to be a letter to you. As of now you need to have read the introduction, to be oriented toward your letter. One point to you is that before I could ever, if ever, get up to your school to talk, I have to finish this manuscript. I'm hoping to do that by the first of October, the first week or so. I'll probably be still working out the cover design. In the meantime, it is not necessary that I come there. What is necessary is that the stronger leaders counsel the frightened and overwhelmed with their particular situations with regard to gasoline and heating. That is a role which the faculty can take with the students, the churches can take with the communities. I've had people tell me it takes time for the government to do things. We don't seem to realize that we can make our wishes known to the government. We needn't be passive. Some people are more experienced with organizing than others. The churches have a sort of edge on that in so far as they have networks. Perhaps the American Public Transportation Association could help us form logistical plans, the local transportation authorities speed up plans, for which provisions have been made. Boston has some increase in funds for Worcester to Boston, and the Southeastern route. The commuter train was under consideration in Central Massachusetts in the transportation bill that was just passed. There are funds for buses, Greyhound told me. Bus companies were sprouting up and dying out in that area. Peter Pan is well established. I do believe that Boston and New York City could do much better with their commuters who drive. I am familiar enough with both locations to have that faith. The commuters need to join the system and the system has to grow for them.

In the long run, kinks worked out, it could be a smooth and brief ride to work and back, with no road rage, no increased blood pressure, no wasted time in gas fumed traffic jams, and more time with the family at the end of the day. I don't know what you do with the trucks on the local highways, but they'd move along better if there were one bus for every 20 cars. [Maybe some of us should go Germany, like the Greyhound public relations man did, so we could see what it's supposed to look like.]

Well, this was the most impersonal letter I ever wrote. You would enjoy the other a lot more, I'm sure. So I'll send that along, too. But I invite you to read my book. Here you are in it.

Yours in Christ,

Mary Ann Segal, M.D.

CHAPTER FOUR:

OTHER OPINIONS ON OIL

OTHER OPINIONS ON THE OIL CRISIS

I have a transcript from a Jim Lehrer News Hour from an interview called THE FUTURE OF OIL, June 1, 2004, where Jeffrey Brown interviews Paul Roberts and Daniel Yergin. Paul Roberts wrote a recent book, entitled *The End of Oil;* he writes frequently about fuel and economics. Yergin, wrote *The Prize: The Epic Quest for Oil, Money, and Power,* which won the Pulitzer Prize. Mr. Yergin is chairman of an international consulting firm that is called Cambridge Energy Research Associates.

Mr. Roberts says that the optimist say you don't hit peak production until 2040 and so there is plenty of time to get ready with alternatives--[at the rate we are moving, that is very soon, as well.] He disagrees, however. He raises the question about whether the market was tight in 2004 or whether prices were driven up by speculation.

[In the *New York Times* article, "As Oil Prices Soar, OPEC Says 'Not Our Fault' ", June 6, 2004. The high prices were attributed to factors over which OPEC had no control. They listed four reasons for the high prices: The first was the buying of futures (speculation); tightness in the U.S. gasoline market; geopolitical concerns; the fourth, higher-than-expected oil demand growth, especially in China and the U.S. At one point, Neela Banerjee, the reporter of the long series of articles in the *New York Times* that began sometime in the spring of 2004, stated that an actual deficit of barrels developed to a small degree, with the demand so high that most of the world's oil companies have been pumping out to top capacity. So Mr. Roberts' concern about speculation was touched upon. I prepared this portion, in 2004 and am editing it now in September, 2005. The reader would be interested to know that on June 6, '04, oil cost about $42 a barrel.] But he feels that we will run into trouble much sooner than the optimists: It is getting harder and harder to find oil—"The rate at which oil companies can find oil and put it on their books for later production...has been declining since 1961. Right now, we burn about 29 billion barrels a year....We're only discovering around nine billion barrels of oil." He expects non-OPEC oil to peak in the next ten to twelve years. Non-OPEC oil is from the North Sea, the Caspian, the European Arctic, and elsewhere. OPEC will peak later, according to Roberts. [Some oil geologists believe it has already peaked, is peaking, and many agree it will be peaking soon. Someone stated, I can't recall whom, that we shall know it has peaked only after the peak. I shall discuss the meaning "peak" when I finish this commentary on the PBS program. (see summary of David Goodstein's *Out of Gas*, Norton, 2004., end of this chapter.)]

Mr. Yergin states that oil production at the end of this decade worldwide will be 20% higher than it was at the beginning of the decade--on a field- by-field analysis.

[I insert my thought here that Exxon-Mobile stated that we will need to get 50% of our oil from new fields and reserves, by the end of the decade. So there is a struggle between these two sets of statistics. If there is the continued growth in demand, that has occurred all along in the past, we shall have a demand that exceeds the supply, resulting in higher prices and a gap in the number of barrels produced with that demand.] Mr. Yergin feels things will be only tight for the next two years. "And right now, even with these increases that OPEC says they're going to do, we're in a very tight situation where the market is vulnerable to significant shocks. We have to be prepared for that."

Paul Roberts disagrees with Daniel Yergin. As he built his case in his book, he sees the end of oil pushing us into: a whole new energy regime. Cheap oil is a thing of the past, but "The good news there is that tends to push the economy into a more efficient behavior," with a reduction in demand through more efficiency and a greater interest in alternative fuels. At this point, sudden volatility or instability in an oil production country is very dangerous, more so than running out.--But he brings up that we also have to address the way we produce our energy, implying we don't have decades to reverse the way we do produce energy *because it is impacting our climate.*

It was then discussed as to whether the United States could take seriously the problem of climate change; and in addition, --what Europe already does-- to have a gas tax. At what point could consumers take this seriously in our country. For the higher price, of $5, has helped the Europeans to be clearly affected in how they use energy. [I think here of the discussion I had with an SUV owner that I met in the Boston area, who with me bemoaned the inefficiency of his vehicle, but stated that in Rome, Italy, he had an Opel station wagon that gets 80 miles to the gallon and adequately holds several passengers, with room for baggage. So frequently, one sees American car ads, now, after we've faced a crisis in oil prices, that offer the same old seductive styling of a "tough" vehicle with a ravenous gasoline appetite. The hybrid cars recently released on the market get about 50 miles to a gallon. The Opel the man owned in Rome was not a hybrid--It was a using computerized fuel injection to maximize burning efficiency. Where we have that in our American cars, it is only a much less effective system. Regrettably, Detroit has not been trying to get fuel efficiency and has been appealing to the consumers with tough looking vehicles that burn like trucks, with no advertised alternatives that offer the space that families need to seat enough passengers. There are about 20 SUV's on my city block. Reasons ? We need the room for the family or I need the room to transport what I carry in my small business.]

The discussion acknowledges that China wants a car economy and will be burning plenty of fuel, but that America could take the lead in developing new technologies to get to the place of turning ourselves into a climate-economy instead of a carbon economy. [Their discussion of alternative fuels makes it clear that they are unaware of the possibilities of a hydrogen economy. They say billions and billions are being spent on research. The solar and wind hydrogen economy researched, and tested to a large degree, which is analyzed by Harry Braun is not his research, but that of many people and many projects over the last century. By being ignorant of it, we are dabbling in many areas and not attuned to a system that could provide us with a complete replacement of fossil fuels and nuclear energy. Now many projects, one of which I communicated with here in New York State, are reinventing the wheel of how to use hydrogen.]

. . . .

On January 15, 2005, *Wall Street Week* on PBS aired a program from *Fortune,* with the content "How to Kick the Oil Habit" by Nicholas Varchaver from the Aug. 23, 2004 article. The subheading is "Gas prices are soaring, pipelines are burning, oil supplies are tight. Here are four ways to fix the mess before the well runs dry." First he enumerates the several incidents that can send "ripples of pain through the stock market, through oil-dependent industries such as airlines, and finally to consumers, who complete the trifecta of suffering when the cost registers at the pump as $2-a-gallon gas." "...U.S. energy security--making sure we have enough oil to run our economy--remains shockingly vulnerable. Production capacity is stretched so thin, demand is so high, and supply is so fraught with uncertainty that we're just a few riots or explosions away from another oil crisis. The lack of spare capacity is eerily similar to what it was in 1973, says Peter Schwartz, a former scenario planner for Shell. That year, ...prices tripled within months, and gas lines became a symbol of national impotence...Indeed, even without a major disaster, some analysts now believe we're likely to see $50-a-barrel oil before we see $30 a barrel again." He goes on to say that in 1973 we imported 30 percent of our oil, whereas now we import 60 percent. We have been on a two decade "oil pig-out, ...the U.S. has increased its already world-leading consumption by 6% in the past decade." And yes, China is using more and more, but we have to look at ourselves. Much as we'd like to, we can't blame it on OPEC.

"High prices and disruptions in supply aren't the only problems. Some veteran observers, like investment banker Matthew Simmons, think we are nearing the point--if we are not already there--at which the world's supply of crude peaks and then begins to decline. (Simmons, by the way, is no ponytail-wearing radical. The 61-year-old Houston oilman has advised the 2000 and 2004 Bush presidential campaigns and Dick Cheney's secret energy task force.) Even the optimists believe the start of the downward slope is only 35 years away. Frankly, it doesn't matter who is right. Three decades is precious little time to reconfigure the world's energy system."

Pulling a string like making oil from the tar in Canada's bountiful tar sands requires using natural gas which he explains is on the rise to less availability too. That, in time, will also be necessarily imported and will be another OPEC situation by 2020, according to Amy Jaffe, associate director of the James A. Baker III Public Policy Institute at Rice University.

The article goes on with many excellent points. First he notes three proposals that have been decades-long standoffs between the two parties. The GOP fights hard for drilling in the Arctic National Wildlife Refuge which might yield a million barrels of oil a day. Democratic hardliners try to get miles-per-gallon requirements for cars increased, which have not been increased substantially in 20 years. And the biggest third rail is trying to raise the gas tax by a dollar or more a gallon to promote a dent in our oil use, but opponents insist that would cause "rioting in the streets." Polls indicate that Americans oppose the last 2:1. Yet what we pay in taxes shows the real cost of oil to be more than $5 a gallon. The peace and war-time costs of Persian Gulf activities are a large part of that cost. Harry Braun and others add to this the public health costs of pollution and the damage to buildings and the environment as additional costs of oil use.

This next is very striking to me: $1.4 trillion dollars is what the U.S. and Canada will spend through 2030 on oil and gas—"everything from exploration to infrastructure such as pipelines and refineries--to handle growing demand and changing sources of supply, according to the International Energy Agency. And that's not even counting the additional $1.7 trillion required for our electricity infrastructure."

I'll end my review of this article for now on the following note, to take it up at some other point with a review of his suggestions, which are not in complete accord with my own, but are in several respects. [He looks for a solution in celluloid rich biofuels, like switch grass, that grows along fields, like a weed. They are potentially more potent ethanols than corn ethanol, he thinks.] What should be noted strongly is that he has delineated a figure which is in significant accord with the amount of money that Harry Braun currently estimates it would cost the U.S. to produce 100% of the energy now used by us with oil, natural gas, coal, and nuclear power--preparing the infrastructure for wind and solar generated liquid hydrogen fuel. *Three trillion dollars.* And, it so happens that regarding electricity, hydrogen fuel can be used in power plants right next door, because there is no pollution, only water. Moreover, the hydrogen pipeline is a superconductor of electricity, no loss of electricity due to resistance, like wire has, for any distance whatsoever. So we could integrate these benefits of making electricity and gridding electricity into revamping our electricity infrastructure. *Mr. Varchaver next states that "no matter what energy reforms we put into place now, there's no way to avoid spending a substantial chunk of this looming $3.1 trillion tab".* But the question for America is clear. How much longer do we want to spend our resources simply patching up a system that funnels money to unreliable foreign despots and is based on a commodity we know is sooner or later going to run out ?

.   .   .

Now, drawing on David Goodstein's *Out of Gas,* [W.W. Norton, 2004], I'd like to help clarify for the reader what "peak" means, with regard to oil.

British Petroleum falls in the "optimist" class. The amount of oil cited by British Petroleum is on their website. According to them, the ratio of known reserves of oil to rate of use: the R/P ratio—is forty years. I am paraphrasing from David Goodstein in *Out of Gas,* [W.W. Norton, 2004], pp. 120 forward. In a BP graph one can see that the amount of oil available in 2001 was approximately one trillion barrels. We have also already used one trillion barrels, since we started burning oil, steadily increasing demand as we went. So the oil, that which we have not used, is still in the ground, waiting to be pumped. Since we're only using 1/40 of that per year, why won't it last another forty years?

Now Goodstein introduces the following factor. Hubbert's peak. This notion was first introduced by an oil geologist of the same name, Hubbert, as it applied to U.S. domestic oil. Realizing that the amount of oil being discovered was gradually declining over a period of time, it led Mr. Hubbert to predict a bell shaped curve, a shape like a hill, but with the back side of the hill curving inward, somewhat. When the peak occurred we would be at the top of the hill, having taken out of the oil supply that which is most readily pumped. Thereafter, the amount retrieved, overwhelmed by the amount used, would result in a deficit of supply, a depleting supply. This prediction, though ridiculed for some time, turned out to be accurate for U.S. domestic oil supplies. In the 1970's we began importing more and more oil.

Applying the same criteria to the world supply, "when the rate of increase of known reserves reaches zero (which for all practical purposes may have already happened), we will for the first time in history be consuming oil faster than we find it." [Goodstein, p. 121]

For the Hubbert's peak predictions [the pessimists] and the BP predictions [the optimists], it "is not a matter of how much oil there is, or of how fast we're using it...The difference is where the crisis will occur. The unstated assumption of the [optimists] is that we'll be fine until every last drop of oil is pumped out of the ground. [Goodstein doesn't say this but many people point out the getting the last half of the oil is technically more and more difficult and expensive, and that the quality of the oil is less desirable. One journalist, in a private conversation said that oil companies are not willing to go through that expense; and I guess we could add that the consumer may not want to buy that expensive, dark bitter crude, which is so difficult to refine. In discussing the oil crisis in 2004, Neele Banerjee, reporter on oil for the *N.Y. Times,* was contrasting the desirable light, sweet crude, with the oil in Saudi Arabia that needs much more refinement.

David Goodstein concludes, "At the halfway point, when falling supply meets continuing rising demand—If we have already consumed nearly half the oil there ever was, the crisis can't be far off. [Goodstein, p. 122]

It seems to David Goodstein and to me and to many others, that we are near enough to the end of oil, not to play Russian Roulette. For those who don't know of this *macabre* game, two players point a pistol, in turn, to their heads, pulling the trigger; the loser is the man who is victim of the one loaded chamber first. Whatever the number of years we can last with oil, I say we should use oil wisely. In Goodstein's words, "We should try to kick the fossil fuel habit altogether, as soon as possible. In 1960, John F. Kennedy challenged us to put a human being on the moon within that decade. And we did it! That was possible because we already knew the basic principles of how it could be done." [Goodstein, p.122]

.   .   .

I think now of WWII, for as a child I was aware of the war and the responsibilities of the fathers, the Victory Gardens, the air raid drills, the rationing, and the news reels. Later I learned that the whole country tooled up with factories in just one year, to build bullets, bombs, and planes. That the women worked the factories, while the eligible men went over seas. And I learned that there were German submarines in Bayonne Bay, in NJ, just a few miles from where I lived. Oil is a world war to me, not that countries are each other's enemies, so much as that oil shortage is a danger to most of the world; to all of it, because we are so interdependent.

Goodstein said we did it, because we knew the basic principles to get to the moon. Harry Braun, exasperated, said of course there is a prototype,   for wind generated electrolysis of water to get hydrogen. His reference was to an interview done by Peter Jennings, years ago, with an elderly farmer in the Mid-West, on Person of the Week. It is available on Harry's video, by the same name as his book, *The Phoenix Project: Shifting from Oil to Hydrogen.* I repeat that the book is a must to read and that the tape is very useful and well composed. However, beware—the interview Peter Jennings does, comes after the credits. The elderly farmer had put up a small, "rinky-dink" windmill, but got all his electricity with it, and went on to get all his own fuel, in the form of hydrogen, with the same windmill. He had a sixth grade education, but was imaginative and mechanical. That is a simple principle, not rocket science. If I were teaching middle school now, as I did in the sixties, the seventh and eighth graders would be learning the simple principles. At first hearing, I didn't catch on to Harry's wind/solar approach. I asked him a very ignorant question, as I attempted to ask about his work, calling him with the contact information I got from WNYC, having heard him call in about the World Trade Center disaster. He mentioned the BMW dual tank cars and he mentioned the Boeing-Lockheed engineering designs for hydrogen fueled jet planes. Hydrogen. That stuck with me. That made sense. But the question I asked was, "How much gasoline does it take to get hydrogen out of water." The wind turbines were not a foregone conclusion for me either at that time. I had to sit with it for awhile.

Harry had a wind farm in the Southwest for a time. He said, insightfully, people have blinders. Very few in the wind business know that they could also be doing electrolysis of water for hydrogen. Over the past two years, I've been urging workers that I call, doing my normal bank, credit card, computer technical help, phone companies, etc. business to be aware that if there were a fuel emergency and they live in a district with wind farms, water, agriculture, and a relatively small population, they could probably use the turbines for some hydrogen, too, in an emergency.

We hear talk of doing research. Then more research. Yet Braun analyzes research done over a century. Since WWII we had completed or started one project after another for hydrogen. But we put it off. Now we can't any longer. If we imagine that we're setting out to research hydrogen, we certainly are ignoring NASA that has used it so much. And we are ignoring the projects so well developed in Mr. Braun's book. Braun, who in turn, refers to Peter Hoffmann, *The Forever Fuel: The Story of Hydrogen,* [Westview Press, Boulder, CO] 1981.

Hydrogen, from splitting water, H_2O, into H_2 and O_2 by electrolysis, is a forever renewable process. Simply because burning hydrogen, recombines it with O_2 and makes water once more. Two-thirds of the world is sea water. There is no shortage. Some techniques of getting hydrogen can be used to get vast amounts of desalinated water, as well, far more inexpensively than current methods. That technique is using Ocean Thermal Exchange Conversion. But let me not confuse the issue. (See Ch.8)

Let's assume that DOE's Hydrogen Posture Plan is correct that wind energy is the cheapest way we know to get hydrogen, by water electrolysis. Turbines turn in the wind, they turn with steam forced upon them in an electricity plant, etc. To get electricity, one must have a generator associated with the turning, the rotation. A generator is made of usually copper wired spools in a magnet's field. The turning of the copper coils around the magnet induces the electrons in the wire to joggle backwards and forwards [because of the positive and negative pole in the magnetic field] to make an alternating current of electricity. That current is sent to wherever you need the electricity. Let's assume that you built a certain turbine, or multi-array of turbines, as in the Heronemus windship (illustrated on the cover). The electrical energy is reduced in making hydrogen, but it is also converted to a fuel which carries energy to be used at your disposal where fuel is useful and electricity won't solve the problem.

In fact, because as a fuel in a combustion chamber making superheated steam, hydrogen is much more efficient than fossil fuel by a factor of 2X's, in generating electricity locally, energy consumed in releasing H2 from water, is regained to a large degree in its power as a fuel to make electricity. [Braun gives the citation for this on pp. 104-105, Ibid.] Why do we worry so much about the expense? There is no other perpetually renewable fuel; and there is no other fuel with the ability to propel a space ship and also make a jeweler's torch to solder platinum and gold. I will show Mr. Braun's cost analysis for hydrogen—with mass production, the cost of a kW of electricity will probably be reduced to one or two cents. The cost of a gallon equivalent of liquid hydrogen would be perhaps, $1.40 in comparison with gasoline at $1.00 a gallon. [From Phoenix Proect website pp.] But gasoline is subsidized directly, and much more, indirectly. So it is never a dollar a gallon. It is more like five dollars a gallon. Some could add up costs to $50/ gallon. To have a fair market, once we could have the hydrogen, we should subsidize the hydrogen, and tax the oil for its real expense. That is not desirable at this moment—The price pinnacle is already breaking some budgets, for gasoline. But we must go ahead with mass produced hydrogen, and shift our fuel use to hydrogen, without the neurosis of worrying about what it costs, to do things properly.

What research does it take to see that hydrogen comes forth from water? I am so pleased with the Hydro-Gen company, whose water tank in the car can serve as an electrolysis site for oxygen rich hydrogen to mix with gasoline and save about 21% of the gasoline. The current is available from the running motor. The electrodes, positive and negative are properly wired and placed in the water. There is an electrolyte in the water, which is a chemical that permits electricity to flow through the water. The positively charge H+ goes to the negative electrode and the negatively charged O= goes to the positive electrode. Thus the H2 becomes free and collectible, though it is not available in the atmosphere – This is simple, not rocket science, but it brings one joy that life can have simpler solutions than wasting lots of fossil fuel to grow lots of corn on lost farmland, subsidize it with three billion dollars a year, and come up with a product that is negative in energy—That is makes less energy than was put into its cultivation, irrigation, harvesting, etc. with fossil fuel, giving off green house gases greater than if the gasoline it replaced were never replaced. For the simple Hydro-Gen go to [savefuel. ca] The cost? $197.

A scientific and energy journalist, in a one on one discussion, told me, "They won't invest unless there is a prototype." To anyone with whom the concepts have been digested over time, the need to *prove* everything becomes an irritant. He told me that the Germans have started a wind turbine, hydrogen prototype this year, and thought he may have heard about it from BMW. He does know of an island off Norway, where three wind turbines supply all the electricity and hydrogen fuel that its inhabitants need.

He agreed, when I suggested that once the concept of electricity from turbines is established in a person's mind, one could turn back to the electric plug to prove many aspects of hydrogen's usefulness. As I said, it is used in "water torches", where precious metal jewelers electrolyze water from the tap, with electricity from the wall plug. The fire is hotter, and there are no impurities to effect the quality of the precious metals being melted. My close friend, a very skilled and artistic jeweler, told me about this. He said he didn't like it, because you can't see the flame. I concluded they must go by the effects of the solder melting, as what they see. Mark said that must be the case.

So really, by making a simple shift in your mind, one could demonstrate hydrogen coming from water, by electrolysis, from the wall plug. Just think, this could be from the generator wires of a wind turbine, or large array of wind turbines—or from photovoltaic cells, but that's a little harder to explain. They cost more, so for now, wind is in the lead. Hydrogen from human and animal waste treatment is quite inexpensive and a necessity, too, for water purification. Electricity can be generated with turbines in ocean currents, too. I don't know what the overall potential is for how much energy one could get from ocean currents. But I did have very interesting discussions about this with Dr. Michael Muller, at Rutgers University. He thought it to be of value to capture the oxygen too, while doing electrolysis of water. It could augment hydrogen release, by ridding the system of the pull of the negatively charged oxygen. I believe he was trying to say that the oxygen would pull at the hydrogen, if it remained nearby. In any event one does wonder why the oxygen is not used, as he suggested, say to burn with methane or biomass. Biomass is mostly wood, and if not scrubbed is a detriment. Of course, methane would have to be captured, "rogue" methane, as from a swamp, wastewater, or landfills. Rogue methane that gets into the atmosphere is more damaging to the greenhouse effect than is CO2.

Now the rocket science crew may again say that nuclear generated electricity is better than wind generated. And then minimize the dangers of nuclear facilities and their waste product dangers, the unsolved problems of how to store and where to store nuclear waste. The amount of fossil fuel in mining and other processes of building nuclear plants is also great. I found a quote in Goodstein's book, who is a proponent of trying further efforts to get nuclear fusion [no container to hold the heat, thus far.] A magnetic field is one thing Goodstein mentions having been suggested. Fusion is what goes on in the heat of the stars, fusing nuclei together to build up the chemical table. Fission is splitting the atom. "However," says Goodstein, "known reserves of uranium are estimated to be enough to supply all of the earth's energy needs—at the current rate of energy consumption—for a period of only five to twenty five years."[p.106]

Goodstein also adds, that to make enough nuclear plants just to meet the needs of our energy requirements now, the largest practical plants would have to be built at the rate of one a day for thirty years, or 10,000 of them. I believe it takes at least ten years to put up one nuclear plant.

Instead of 10,000 nuclear plants, let's quickly manufacture one million Heronemus windships and get the 100% fossil and nuclear fuel replacement that it would take for the U.S. One million would fuel current U.S. consumption. This is based on Harry Braun's analysis at his website, [phoenixprojectfoundation.us] Harry estimated they might cost three million apiece, if mass produced. He was explaining to me that they could be manufactured from the engineer's design to hardware, in three months, like a certain fighter plane was during WWII.

It has just occurred to me that *four* million *windships* would fuel the entire world at current consumption rates. If it is correct that land turbines cost proportionately more, the world could have wind ships for what is now used throughout the world, of energy, for the same cost it would take to supply just the U.S. use, with enough land wind turbines, at one and a half million dollars apiece. [That is, ten million one megawatt wind turbines with which we are familiar, at ten to fifteen trillion. Land turbines then would cost about the same or more than four X 3trillion, or twelve trillion for the U.S. alone. Twelve trillion for four million sea windships could supply the world's current use of energy sources to be replaced, all clean and forever renewable.]

The windships are pictured on the cover of my book. They are 500ft. tall and 250ft. wide, have thirty six huge turbines, get much stronger wind at sea, and get 10-18 MW apiece. They do not take up the entire coastline as I first pictured them. One million, say split between the Atlantic and Pacific, and properly placed so that they didn't block the wind from one another, would take up 640 square miles, still in the shallow coastal plain. That, split between two coasts, would be bit more than 12.5 miles by 12.5 miles. They need not be in sight of land, though they would be quite an exciting sight. Mr. Braun figures them to be from 5-15 miles offshore. Harry has been analyzing the ways in which to get hydrogen from large projects that have been funded over the decades, many decades. Heronemus' windship is his favorite choice. I favor it too, and had the good fortune to locate the woman who has her father's patent rights. She said they have made some improvements on her father's design.

If NanoSolar, Inc in California is right about its potential, solar energy would cost 1% of the cost of photovoltaic cells and be 5X's as efficient. If this is a genuine breakthrough, infrastructure would be very inexpensive throughout the sun hotspots in the entire world. I know that they are printed out with nanotechnology. In one issue of the *Journal of the International Hydrogen Association,* researchers were looking for polycrystalline material, to be an improvement over the single crystal, silicon, in Photovoltaic Cells. The expenses mentioned however, were appalling, compared to the wind turbine generation costs. This nanoprinting of plastic crystals in three dimensions mentioned by NanoSolar, Inc. may be the way to release magnitudes of more energy. [See notes on ch.8, where Heliovolt's thin film cells, made of copper indium selenide, are showing great promise.]

I personally do not see the need to use more and more energy, beyond providing the world with necessities, and would consider for mobility very fuel efficient hydrogen cars, trucks, buses, trains, motor bikes, bikes, with efficient public transportation planning, as necessities. That's at least until the two billion with no energy whatsoever, not even to cook, or study in the darkness, are provided with energy, before everyone drives a car, throughout this world. There are billions who still want cars. What would that do to keeping up with hydrogen fuel? Let's think about it this time, before we start up a wild machine, that has no brakes.

Though hydrogen is forever renewable and there are magnitudes of energy more than we currently use, it is wise to decipher what we are doing with it, at what rate. Solar electricity for hydrogen has not reached its state of ideal development, so some of the magnitudes more energy are not yet available.

CHAPTER FIVE:

ARE ETHANOLS AND BIODIESELS AN ANSWER TO REPLACING GASOLINE OR DIESEL FUELS?

ARE ETHANOLS AND BIODIESELS AN ANSWER
TO REPLACING GASOLINE OR DIESEL FUEL?

The United States, and the rest of the world, desperately need a liquid fuel replacement for oil in the future. In Chapter Three, I listed the top list of solutions that are being promulgated as answers, or partial answers to the shortage of oil products. E85, or ethanol 85%/gasoline 15% was first on the list mentioned in the *New York Times,* 9/10/05. It was selling for 40 to 50 cents less than gasoline on the day it was reported. Burning vegetable oil as a biodiesel fuel was also on the list of four or five alternatives. E85 is corn alcohol, produced in the Midwest, primarily.

My first reaction to hearing about ethanol enthusiasm was simple, when Howard Dean was running for president in 2003, and his energy chairman was committed to this fuel as a solution, as was Mr. Bush, [and possibly John Kerry—I can't recall his position on ethanol]: How can one grow fuel for modern civilization's energy needs? The fuel crop, widely used, would take up land and water that is absolutely necessary for growing food.

My friend Mark argued that food alcohols, for example, potato alcohol, were used by the German farmers during the war, NASCAR uses alcohol, and the amount of carbon dioxide coming out in the burning process, is the same amount that the plant took in. So there is no net change in CO2. I know that decaying plants give off CO2, so what's the difference?

One difference is that with modern farming, a lot fossil fuel goes into ethanol's growth, irrigation, harvesting and distillation. And don't forget that the fresh water supply, due to ground water depletion, is lost for drinking and for food agriculture, if it goes into irrigating the ethanol crops. I had heard that America's grain storage is the lowest it has ever been; and later I read that for four years now, the world's people [who ate] have eaten more food than was grown.

"Ethanol can be produced from a wide variety of plant-based feedstocks, most commonly grain or sugar crops. It is then blended with gasoline as an oxygenate or fuel extender for use in gasoline vehicles, or it can be used alone in 'flexible-fuel vehicles' that run on any blend of ethanol and gasoline." [*Ethanol's Potential: Looking Beyond Corn,* Danielle Murray, *Eco-Economy Updates,* June 29, 2005, Earth Policy Institute]

Though the next paragraph in this new update praises ethanol as an emission reducer, including emissions of fine particulates and carbon monoxide, and improving rural economies by supplying jobs and higher farm incomes, the following paragraph states the "today's inefficient production methods and conversion technologies mean that this fuel will only produce modest environmental and economic benefits and *could impinge on international food security.* [Italics mine] The largest obstacle to biofuel production is land availability. Expanding cropland for energy production will likely worsen the already intense competition for land between agriculture, forests, and urban sprawl." The author goes on to say that food security is precariously balanced with temperatures rising, globally, and water tables [underground water] falling worldwide.

Ethanol needs to be made with an emphasis on land efficiency, I paraphrase. Corn, with broad political support in the U.S., "is one of the least efficient sources of ethanol." French sugar beets and Brazilian sugarcane, she states, are twice as efficient in yields per acre as is American corn.

"Growing, transporting, and distilling corn to make a gallon of ethanol uses almost as much energy as is contained in the ethanol itself." Here, it is said that sugar beets produce nearly two units of energy for one unit used to make it. Sugarcane yields <u>eight times as much energy</u> as needed to produce it. So sugar crops are more efficient than grain crops.

It is observed that sugar crops could widely increase in many tropical countries, cutting oil imports and boosting rural economies. "Unfortunately, new fields may cut further into already shrinking rainforests, making them a serious environmental liability." [I want to comment here, that one needs to worry about whether the rural and urban poor are fed. There's sort of a sweep of logic, unless the author means that the wide increase is *apt to occur*, rather than that she is recommending that it occur.] No, she isn't recommending it—but the reader and I might have thought so. The next paragraph states that "If ethanol is to become a major part of the world fuel supply without competing with food and forests, its primary source will not be grains or even sugar crops", but rather cellulosic plants, like agricultural residues, forest residues, grasses, and fast growing trees. [If fast growing trees can be used for wood to protect older forests from being cut down, and by that I mean wood for furniture or floors, etc. that's a primary need. To burn wood mass per se, is to put nearly 200 chemicals in the air, including some carcinogens. Methyl alcohol, that is wood alcohol, is poison to ingest, but in burning it, there is complete combustion, so it emits only CO_2 when burned.]

Switch grass is the most up to date in the thinking about potential sources of ethanol, I believe. Switch grass was recommended by Mr. Varchaver, of *Fortune,* whose ideas on oil shortage are discussed in Chapter Three. I mention the switch grass now but not much in that chapter. But if you can remember, it was he who gives the figure of 3.1 trillion dollars for any energy system to be put in place, between the U.S. and Canada. In Danielle Murray's article which I've been drawing upon here, she states that switch grass is used by farmers to prevent land erosion. "It requires minimal irrigation, fertilizer, or herbicides but yields 2-3 times more ethanol per acre than corn does. Such crops could potentially be harvested on marginal land, avoiding the conversion of healthy cropland or forests to energy-crop production."

[I would like to point out that getting clarification of all of this is very important, for governments should not allow corporations or individual farmers to be using necessary croplands for profit from the better-off who would buy ethanol from them, regardless of loss of cropland or forestland. As it stands now, governments are subsidizing ethanol production. Danielle Murray believes that there is little net energy from the corn ethanol grown in the U.S. Next, I'll review the findings of Pimentel and Patzek in the *Natural Resources Research* journal, from the faculties of Cornell University and University of California, Berkeley, respectively. Their study indicates that there is <u>a net *loss*</u> of energy in growing corn ethanol in the U.S.]

Ms. Murray concludes by saying research and development is needed to improve biomass-ethanol conversion technologies. She states that biofuels will only become more than a fraction of our fuel supply, if fuel-economy standards aren't made for our vehicles, or if plug-in hybrids aren't used to draw on electricity, preferably made with wind. In this way, perhaps fuel demand will be reduced to what ethanols can provide.

.   .   .

A wind and hydrogen colleague of mine alerted me to an article about the production of bio-fuels consuming more energy than it produces. I had not had a great deal of interest in ethanol, having heard that it did not burn smoothly in the motor of an acquaintance who was required to use the mix in New Jersey. As I thought of the competition between food needs and water needs and the large scale growth of fuel crops, which if widely used as they are supported here in the U.S. would require all of our cropland, by one estimate, and twice the land mass of the U.S., by another estimate.

However, I was very interested in the issue, raised by David Goodstein, in *Out of Gas,* cited for his excellent analysis of the difference between the oil peak analysts and the R/P analysts, who say one can go down to the last barrel in the earth, by their statistics. [See Chapter Four]. The important issue that David Goodstein raises is the issue of the decay of plants. It is completely natural and inevitable in the simple world of nature, and was taken for granted, before the accumulation of a lot of CO_2 in our atmosphere that is the result of more than a century of burning fossil fuels at ever faster rates. But there are many plants that die and decay. They give off CO_2.

David Goodstein raised the question of whether we should, now, advisably bury plants that have died. I wondered if more carbon were in the ethanols than in the decaying plants—do decay and combustion produce the same amount of carbon dioxide? I presume that the answer is yes, there is a given amount of carbon in the plant and it makes CO_2 whether decaying on the ground or being burned in a biofuel. That was the question I was asking a few colleagues, but no one had thought about the answer. It was basic, very basic biochemistry, but too simple for any of our overloaded minds that week. I know that no carbon is added in the distillation to alcohols. However, the amount of fossil fuel and CO_2 and other impurities emitted in the growth and processing of biofuels is a very big issue.

An issue so heretofore unthought-of as "Do we need to bury dead plants and their fruit" has now been raised because of the great overload of carbon dioxide in our atmosphere. Blocking my mind from a thorough ability to conceptualize plant sequestration under the soil is the notion that it should even have to enter our minds.

Anyway, what I was getting at, is that I didn't know what Dr. Pimentel's work was about, nor how to get the journal article. But I did have his email address, so wrote a letter, explaining what I am doing, and I believe I asked him my question, about the equality of CO_2 emission burning plants as in plant decay. We never landed on that answer to that question, exactly. But he answered many others that I raised.

I'm going to include them in this chapter.

Wed. Aug 10, 2005
To: Mary Ann Segal
From: David Pimentel

Dear Dr. Segal:

Thank you for your letter and interest in bio-fuels. Hydrogen as a fuel, as you indicate, has potential, but it does have its problems.

I have been invited to Brazil four times related to ethanol. How do the Brazilians produce ethanol from sugarcane? Heavy subsidies, the same reason that ethanol is being produced in the U.S. The subsidy for ethanol in Brazil is 50% of the cost of production. Also, sugarcane is a better crop for producing ethanol than corn. However, the use of the *bagasse* as fuel has devastating impacts on the environment.

The production of ethanol in Brazil is also having an impact on food production. It is difficult to find out how much. There is a heavy use of labor in sugarcane production in Brazil.

You are correct, ADM and other large ethanol producers get most of the $3 billion in ethanol subsidies. The farmers at most are getting 2 cents more per bushel of corn. A question I ask, if corn ethanol is so great, why do we subsidize it at $3 billion per year?

Because we use more fossil energy to produce a gallon of ethanol than we obtain in ethanol fuel, we are contributing to the CO_2 releases into the atmosphere. Another major contribution of CO_2 release is related to the tilling of the soil. The organic matter in the soil is exposed during tilling and it oxidizes, releasing more CO_2. See the papers of Dr. Rattan Lal of Ohio State University.

World grain production per capita has been declining continuously for the past 20 years and grains make up 80% of world food. (See FAO). Also the World Health Organization reports there are 3.7 billion humans who are currently malnourished—the largest number ever in history!

Best wishes with your book,
David Pimentel, PhD.
        Ecology and Agricultural Sciences, Cornell University

.   .   .

I guess I answered Dr. David Pimentel's letter that same day, as I had a reply from him on Aug. 11. I recall that I mentioned to him that I spoke with someone from the Institute for Analysis of Global Security, who had advised the Chinese not to get embroiled in OPEC and other oil dependence, like the United States has. He advised them not to burn their rice straw in the fields, but to make biofuel of it and be less dependent on oil. I also must have asked how cellulosic plants are fermented. Cellulose is a fiber cows can digest, but humans can't. We don't try to eat boiled grass. We eat greens without the tough cellulose in them. [I heard from a colleague that a very soft fabric is being experimentally produced from cellulosic material, probably prompted by the thought that we may become short on the products of oil that permit us to make nylon and other synthetic fabrics.]

.   .   .

Thursday, Aug. 11, 2005

Dear Mary Ann Segal:

    The Chinese are strongly discouraging the burning of rice and other crop residues. They also were encouraging the farmers to leave crop residues on the land and not harvest them and then composting the residues. The reason is that removing the crop residues left the land unprotected and intensified erosion 10 fold or more. If you increase soil erosion, you also increase the release of CO_2 and degrade the land.

    If crop residues are treated with an acid and the sugars and starches released, the material can be fermented. However, it does not produce net energy.

    Crop residues decay rapidly if in the soil and if there is moisture. They will also decay fast if there is an abundance of nitrogen. Thatch does not decay rapidly because the thatch remains relatively dry. [I had asked Dr. Pimentel about the decay of cellulosic material, like the grass with which Irishmen and others, in old pictures, have thickly matted on their roofs, for roofing. That seemed not to decay quickly.]

You are aware that biomass wood provides the nation with about 3% of its energy and this is equal to the energy produced by all hydropower. The energy from the biomass is thermal energy, not liquid fuel.

Best wishes,
David

.   .   .

Confused by what Dr. Pimentel meant by biomass wood, I asked him to clarify the statement that I read that Denver, CO has bad air days, the worst air days, when there is a lot of wood smoke in the air. Fortunately, my days with wood burning in the fireplace, as a child, were in a very rural area, with a collection of five houses, and then none for a half a mile, another mile, and then a small village began at the end of the road, approaching the stores and churches and school in the village. The wood burning was sparse. I only remember it in my section of the road, among two or three of the five houses. A boyfriend from high school, now in FL, told me that in 1979, the government issued him a stove, during the fuel crisis that was so developed, that it took less wood than one would expect to heat a nine-room country home, in the same area of western Massachusetts' Berkshire Hills, where I grew up. It was also constructed so that there was very little pollution in the smoke. I never thought of wood smoke as polluting; I guess many a person would be disappointed to hear that, with their nostalgia for camping and country living. It's smell might represent, as it does to me, the association of family, and friendly neighbors, roasted chestnuts, and toasted marshmallows. But the fact is. . .

Aug. 13, 2005

Dear Mary Ann:
    Wood burning does produce a large number of chemical pollutants, about 200 different chemicals and several are carcinogenic. However, in large furnaces most of the pollutants can be removed, like we do with coal. The private homeowner generally cannot do this. This is

the reason some communities like Aspen, Colorado have banned the burning of wood. I am sure that we will continue, though, to burn wood biomass as an energy source.

You are welcome to use my letters in your book.

Best wishes,
David

[I spoke to Dr. Pimentel today (9/27/05) who said that the burning of biomass for thermal energy or heat generally refers to wood biomass. If scrubbed, as in a plant using wood, the results are good, like the scrubbers for coal would be. "At this point we may have to use everything we have!" referring to the fuel crisis. But of course we should convert to clean, renewable energy]

.   .   .

Well, there were a few more letters, but let me get down to the brass tacks of trying to summarize the paper that Dr. Pimentel, Professor of Ecology and Agricultural Sciences, Cornell University, Ithaca, NY published in cooperation with Dr. Tad W. Patzek, from the Department of Civil and Environmental Engineering, University of California, Berkeley, CA.

Okay. I'll try to make this easy. The article is in *Natural Resources Research, Vol. 15, No. 1, March 2005.* The title is "Ethanol Production Using Corn, Switchgrass, and Wood: Biodiesel Production Using Soybean and Sunflower," [pp. 65-75]. The Abstract is as follows:

"Energy outputs from ethanol produced using corn, switchgrass, and wood biomass were each less than the respective fossil energy inputs. The same was true for producing biodiesel using soybeans and sunflower, however, the energy cost for producing soybean biodiesel was only slightly negative compared with ethanol production. Findings in terms of energy outputs compared with the energy inputs were:

Ethanol production using corn grain required 29% more fossil energy than the ethanol fuel produced.

Ethanol production using switchgrass required 50% more fossil energy than the ethanol fuel produced.

Ethanol production using wood biomass required 57% more fossil energy than the ethanol fuel produced.

Biodiesel production using soybean required 27% more fossil energy than the biodiesel fuel produced. (Note, the energy yield from soy oil per hectacre is far lower than the ethanol yield from corn.)

Biodiesel production using sunflower required 118% more fossil energy than the biodiesel fuel produced."

In the body of the paper, the two researchers point out that there were two panel studies by the U.S. Department of Energy in 1980 and 1981, reporting a negative return in producing ethanol and liquid fuels from biomass. The findings were reviewed by 26 expert scientists independent of the DOE. There was unanimous agreement that energy was lost, not gained, in producing corn ethanol. Other investigations, numerous, confirmed these findings over the past two decades.

"A review of the reports that indicate that corn ethanol production provides a positive return indicates that many inputs were omitted (Pimentel, 2003). It is disappointing that many of the inputs were omitted because this misleads U.S. policy makers and the public."

Shapouri et al (2002, 2004) of the USDA "claims that ethanol production provides a net energy return. In addition, some large corporations, including Archer, Daniels, Midland (McCain, 2003), support the production of ethanol using corn and are making huge profits from ethanol production, which is subsidized by federal and state governments. Some politicians" mistakenly believe that its production benefits farmers, whereas the farmer profits minimally.

"In contrast to the USDA, numerous scientific studies have concluded that ethanol production does not provide a net energy balance, that ethanol is not a renewable energy source, is not an economical fuel, and its production and use contribute to air, water, and soil pollution and global warming." He lists several sources, written between 1989 and 2004. "Growing large amounts of corn necessary for ethanol production occupies cropland suitable for food production and raises serious ethical issues."

Dr. Pimentel then goes on to examine the energy output for corn ethanol, in great detail, and then for switchgrass, and wood. Biodiesel production using soybeans and sunflowers are also examined in detail. His figures and analysis in this paper demonstrate his claims and his critique of the findings of their being helpful for energy and the economy, etc. are supported by his figures. His Conclusion gives several physical and chemical factors that limit the production of these liquid fuels.

I'm going to pick out one of the three reasons, which is perhaps the simplest: In ethanol production, the alcohol is made from the carbohydrates by microbes, that on the average bring the concentration of ethanol to 8% in a broth with 92% water. Large amounts of fossil fuel are required to remove the 8% alcohol from the 92% water.

So after the farming losses of fossil fuel energy, the energy for irrigation, for harvesting, the distillation takes more fossil fuel.

In biodiesel production, there is low crop yield per acre for sunflower and soybeans. Furthermore, they are only 25.5% oil for sunflowers and only 18% for soybeans. Oil extraction processes for all oil crops is highly energy intensive.

Another physical and chemical factor which limit the production of liquid fuels such as ethanol and biodiesel using plant biomass materials is that "an extremely low fraction of the sunlight reaching America is captured by plants. On the average the sunlight captured by plants is only about 0.1%, [with corn producing one quarter of a percent]. Those low values are in contrast to photovoltaics that capture 10% or more sunlight, or approximately 100-fold more sunlight than plant biomass."

.   .   .

My conclusion is, burn what's in stock for ethanol fuel, but don't order more. Used french-fry oil, if you can get it inexpensively, doesn't seem to cause any energy loss, and it smells like a fried dinner is cooking, when the car goes down the road, burning it. This information comes from a friend in Florida who takes a great interest in cars. I think you need the right kind of motor to burn biodiesel fuel, so check before you use it! Please note that biodiesel fuel that is new for burning should *not be used,* as its production uses a great deal of fossil fuel and there is already more fossil fuel expenditure making it than it releases in your car. You would not be using an alternate fuel that spares the expenditure of fossil fuels.

.   .   .

After reading this chapter, Dr. Pimentel sent a few additional comments:

Oct. 3, 2005

Dear Mary Ann:

Thanks for the material you sent me. I looked over the ethanol piece and my few comments are:

Page 1, paragraph one: The reason that ethanol was selling for less than gasoline is the large subsidies for ethanol and has little to do with the cost of ethanol production.

1-4: Corn production requires large inputs of water—500,000 gallons of water per acre during the growing season of three months.

1-5: The use of food for fuel does impinge on food production. The World Health Organization reports there are 3.7 billion humans malnourished in the world.

1-7: It takes 30% more fossil energy to produce a gallon of ethanol than is obtained in ethanol.

2-3: It takes about 50% more fossil energy to produce a gallon of ethanol than you obtain from ethanol using switch grass.

2-5: The trouble with the conversion of biomass into ethanol is that the yeasts [for brewing alcohol] cannot survive in much above 10% ethanol. It would be nice if they would live in 90% ethanol.

Best regards,
David

Author's Note: I know that it is distressing, if you believe that something is a solution as an alternative fuel, to have your conviction contradicted. One can be temporarily depressed and confused, by reading a contradiction. Let me say this much, to offset the sense of loss. Dr. Pimentel is Professor of Ecology and Agricultural Sciences at Cornell. He is familiar with information such as the fact that the water under the ground, in the water table, is very diminished in the world. Drawing on that good water for drinking, [that's from an artesian well] becomes harder. Using it for irrigation is harder. I believe that it is the state of Nebraska that has little groundwater left and, dependent on rainwater, is subject to drought. The Midwest has suffered drought and grain output has been difficult just to cover food problems, when the rest of the world is taken into account. *For the first time in recent history, the* world *has produced less food than the world's population has consumed for four consecutive years. Grain storage is at its lowest in its recorded recent history.* If you think about that problem, alone, you would hesitate to use ethanol.

Go to [dp18@cornell.edu] for a copy of the Pimentel/Patzek journal article on biofuels.

CHAPTER SIX:

TAKING STOCK

TAKING STOCK

I want to take a break from the writing and relax and let the reader relax too.

We have Katrina from the recent past and Rita close on the way. It doesn't look good with Rita, either. Naturally, we are very concerned. It's unbelievable how the problems have mounted, doubled, tripled each other.

All the human loss and grueling hardship; all the suffering. And now we have to be careful, very careful with ourselves this winter. I think we have to be prepared to set up heat shelters—There are many areas of our country [and others] where it gets brutally cold in winter. Here in New York it is raw and windy, but is not as profoundly cold as it was in Massachusetts, where I grew up. Yet it is bad enough. But I know that in other regions of the country, people may well be in danger.

With regard to transportation, I know there are programs to take people from a rural area, to the city where they would shop. I don't think that this is strictly for the elderly, though I think there are programs for them, too. However, it would seem that an unemployed but qualified driver could rent a bus, if that size is needed, or use a large van for this purpose. It might also serve to get people to the area of highest employment, locally, at business hours. The cost could be split as fees among the hopefully thirty-five passengers, and be taken as an income tax deduction, if that becomes allowable. It can be requested—if enough people want to claim it, perhaps it will be acceptable. This is in lieu of being able to get a cash incentive as the federal employees do in Washington, for using public transportation. I have left a voicemail with a representative of the American Public Transportation Association to inquire about how that organization might possibly be of assistance. They would know about funding that is already available and just needs to be applied for. They claim a 50% savings on fuel, when a person uses public transportation.

And then, there should be some communication going on with China too. Kenneth Greenspan has disparaged buying of foreign merchandise in an August news article. It was the same day that he warned people who are spending a lot, because they feel the value of their home is great, that the housing bubble is breaking, and these individuals may advisably save more money.

I've mentioned this before, in chapter two, but just touched upon it. Look around the store and notice what merchandise you buy that comes a long distance. Fijii artesian well water comes from half way around the world. Poland Spring water comes from Maine. I was told, but the source was not reliable, that some people were caught bottling Poland Spring water in the Catskills of New York State, much closer to New York City. There was a news report on my computer banner headlines that said 50% of bottled water comes from municipal water supplies. One city sold bottles to put the water in from the tap. We are perhaps afraid of chemical pollution from our tap water. There's a fair amount of unnecessary suspicion of tap water. I suppose the tap water could be tested. It takes a lot of diesel fuel to supply people all over the country with heavy gallons of water. Try to inquire into your local water quality more thoughtfully. Try to buy water, if you don't want the chlorine, or fluorine, that comes from the nearest water supply that is fresh and safe. New York City has Catskill water, some of the best water in the world.

If you are a wine drinker closer to New York, why not buy wine from New York State. There is a trend to California wine, now. Supposedly, it is cheaper, I heard. The cost is probably trivially different and the ride from California is long in a trailer truck. Try to eat local fruit, in season. Trucking fuel has grown faster than any other sector of highway fuel.

I just got a news brief from my best friend, another woman doctor, in California. There was a recent *New York Times* article that indicated that the Chinese may be ahead of us on alternative energy techniques and that they have many scientists and engineers. It confirmed what I read on a Chinese website [china.org.cn], that the Chinese have large areas of wind availability and many turbines—more than any other country. I had to clarify to my friend to think past wind to hydrogen. Electricity does not run planes and steel mills. Where there is wind there is hydrogen. We speculated about the possibility that China may get wind hydrogen before we do. BMW's U.S. representative, David Buchko, with whom I have been having a lot of communication gave me BMW media information that said that BMW has brought a lot of information about their cars and the possibilities of hydrogen to the Chinese for their education. It was translated into Mandarin and provided to all the universities and to an internet system, as well. This exchange of information began in 2000.

In *Out of Oil,* Paul Roberts reports of a trip he took to China. He said they just smiled when he brought up alternative energy, as though to say what is the United States doing about it. However, he did notice some Danish wind turbine company trucks. He was not aware of their massive development of wind. My friend Geraldine is also of the opinion that the Chinese have enough money to move ahead with hydrogen, if they are clear about it. They also have a tremendous labor force. And their labor force is working at a fraction of what ours is. So money may not be an issue for them. And I don't think it's an issue for us, if the banks could put 1.7 trillion dollars into construction of new houses, this year.

I don't know whether I've mentioned this yet, but most of our hydrogen for NASA and fuel cell electricity production is coming from methane and is costing more that way, than it would if wind generated. Methane is not as cheap as when those figures were generated of $2.50 plus, as compared with a $1.00 price for gasoline that I read about. Harry Braun made the calculations for mass produced hydrogen being $1.40. Of course, with methane, there is carbon dioxide left over, too. But do remember that the 1.00 rate of gasoline, would really be much more because of subsidies and external costs, for health problems, military protection, damage to buildings, bridges, crops and forests, etc. [from acid rain.]

I finally spoke to my landlord, the gentleman who is in partnership with the man who is the lawyer. This landlord is a construction chief and does the supervision of the work in all the affiliated buildings. I proposed the savings of oil for eight hours a day, five days a week. He first said that the thermostat outside would make that impossible, but then added that it could be done manually. His next thought was about frozen pipes at 19 degrees, but I said that it doesn't have to get that cold in the building. Forty-five or fifty degrees would save oil. Perhaps a second thermostat could help keep minimal heat. If the temperature were 70 degrees at the beginning of the eight daytime hours, I doubt the building could lose that much heat to go down to 19 degrees in that period of time. There is no pass-along law for fuel at this time, so thinking of his expenditures helped him visualize this with me. He agreed that in exceptional cases a space heater could be used for one room, paid for out of saved oil funds. He mentioned the law, and I said we could both have our lawyers, and there could be an emergency provision in the law for this effort. He could picture the savings in thirty buildings on the block. His last news was that fuel oil would be $4.60 a gallon. [Stated on 9/23/05.] When I said what we save could help Canada or Maine, he was responsive, catching the spirit of the process. So I told him to think about it, and he said he would. We didn't try to think of how much oil it would save in one building. While there is this much activity, a committee could be helping out the people in buildings who chronically go without heat, and working out a shelter arrangement for cases that just couldn't be helped.

[This idea may not work for the following reason: two people have stated to me that they have been informed that a building vacated to go to work should set the thermostat only 10 degrees lower than the nighttime comfort zone—It was stated that reheating more than the ten degrees used more energy than leaving the thermostat higher. There is some savings but is it enough? Will we need to consider a heat shelter system for New Yorkers, too, that is more stringent than the plan I first proposed, if the landlords pass along the heat surcharge by law? The result could be an inability for many to afford the surcharge and the projected raise in electricity costs.]

One acquaintance of mine has been anticipating problems of heating in his Westchester town. He told me that he knows of an oil company that could give the churches, who housed the community centers for such a project, a discount on their oil.

.   .   .

I noted in the chapter, "Other Opinions on Oil", that since we use 25% of the world's oil; and since one million wind ships would cost roughly 25% of what the necessary number of land turbines would cost to do the job of making a 100% of the hydrogen needed by the U.S. to replace all fossil and nuclear fuel;--the wind ship choice could supply the whole world with the energy it currently consumes for the same cost as putting up 10 million land turbines in the U.S. [See Harry Braun's calculations about how many turbines are needed for creating enough hydrogen to replace all fossil and nuclear fuel in the U.S. at [*pheonixprojectfoundation.us* .]

I have made contact with the woman who owns the patents on the windships. She was not able to elucidate too much about their detail, which frustrated my curiosity, of course, but she said that they have made some improvements over W. E. Heronemus' original plans, he the naval architect and engineer who invented the wind ship for hydrogen production and planned to see them off the shores of New England. This woman has engineers working with her, very familiar with all immediately related aspects of the hydrogen production, the basing in the ocean, floating, and simply tethered to the bottom. To carry out this project, she would of course, have to be dealt with directly, so I am very grateful that I was put in contact with her, by someone in environmental engineering at the University of Massachusetts, in Amherst, MA.

Another value to the sea route is that the salt water is used to ultimately gain pure water when the hydrogen is used on land. It can be condensed and captured, and in a motor for just hydrogen, the water would be drinkable. Dr. David Viccarri, at Steven's Institute of Technology assured me that the water vapor from hydrogen would not determine the air's humidity significantly. It would only be approximately double the water vapor we get from burning gasoline, and that would be insignificant in relation to nature's much more sizeable dynamics of humidity and related factors. He was concerned about birds being killed by land turbines I guess, as I had not told him about the wind ship. Since that time, I learned that more birds are killed flying into other human structure than they are by flying into land turbines. But I hope the wind ships work well enough so that 10,000,000 land turbines would not be needed.

Many people say that we should have all useful methods for alternative energy and energy efficiency, and I guess we will as we are all thinking about it. But we see, that Pimentel and Patzek have done the calculations of the amount of fossil fuel used, and polluting while used, before ethanol is available, or biodiesel is available. We come out with a net loss of energy and a lot of fossil fuel wasted, a lot of cropland, shall we say misused. Food is our first priority for the cropland. What we eat can be used to capture hydrogen in waste treatment. What the animals eat, too. This is needed in our country still, and in countries throughout the world. It is relatively inexpensive to get hydrogen from waste and from landfills.

What other alternatives are available—well, Dr. Michel Muller has done a lot of work with turbines in ocean currents to produce electricity. If it didn't interfere with other uses of the ocean, it is good for coastal electricity. Solar techniques are improving, even photovoltaics in greater production. The electricity tape is useful in an army camp. I don't know its cost, but I look forward to the day that something is inexpensive enough, or plentiful enough to supply the two billion people of the world that have no form of energy, and live their lives around getting cooking fuel, and cannot do much after dark, like get an education. I await to hear how NanoSolar, Inc. progresses and truly hope it is a very inexpensive and quite efficient technology. There are other solar projects being perfected, too. Besides Heliovolt (see Notes on Ch.8), NREL has called the Nevada solar electricity project, "near term" to compete with fossil fuels.

People interested in wind electricity are often blind to hydrogen generation with it, not thinking ahead to the fact that we need a liquid fuel desperately. I hope we never have to return to clipper ships to visit around the world. They might be fun, but not very quick. World travel probably helps us to appreciate one another more, and lots of people want to visit their original homes.

Whole towns have been built around photovoltaics and though not cheap and only 10% efficient, they work. [Perhaps their efficiency has grown some, as I read in other reports.] One journalist told me they are 15-20% efficient now. People spend their money for clean heat and electricity. Buying fuel cells, that still use "dirty" hydrogen from fossil fuel, is much more expensive than photovoltaics for electricity generation, if one expects modest results from the photovoltaics.

NanoSolar, Inc.'s website doesn't elucidate too much, [nanosolar.com]. I heard the following interpretation of what they may achieve, a cost of 1% of the cost of photovoltaics, with five times the efficiency. They are making microscopic prints of crystals from plastics, coated with a semi-conductor for electricity production. If that could be sturdy enough, it might release electrical energy and hydrogen in the many solar intense regions of the world. Some oil is necessary for their manufacture, however, as plastic is an oil product. To have some remaining oil would help us continue to produce some medicines, lubricants, and other products. Plastics are piled up in garbage dumps and never go away. dirtying the world, but in some applications they save us wood. In this application, they may be the ideal polycrystalline material that one *International Journal of Hydrogen Energy* issue [April, 2005] recently dedicated itself to. Solar captured energy and wind captured, could give us magnitudes of the energy we now use in the world.

But right now, before we know what becomes of nanosolar techniques, the wind ships seem like the best investment, modules at a time being tested, then a large push ahead to making available a clean, renewable, replacement fuel for any oil product burning equipment.—We need to do it while we still have oil and, in doing so we shall have rapidly achieved energy security, leap frogging the technologies not available for some time anyway, like adequate Liquid Natural Gas [LNG], from other nations while domestic natural gas wanes. Coal gasification is 10-20 years away according to *Fortune*, 2/21/05, with carbon sequestration of billions of tons of CO2 at great cost [and probably, consequences.] [The article was written by Jeremy Main, "Old King Coal is Back".]

ANWR, if started, would take a half a trillion to complete, and there wouldn't be a barrel for ten years, while we'd be in a period of energy gap. It is controversial as to how much fuel is there…change course, folks. It will be an adventure. Why must we beat a dead horse? What is there to lose, one's sense of direction and involvement? Change careers, as you won't be in poverty doing it. Learn how to really laugh and have fun. Celebrate what you've done for the world, question whether it was all good. Retire long enough to think of what would be next to do.

Nuclear plants take ten years to build, and more years to close down when they are corroding. It would take ten thousand of them to keep up with modern demand. Nuclear fuel would deplete rapidly. [See chapter four, with David Goodstein's comments.]

As we put up the hydrogen infrastructure, we'd be able to get to each vehicle on the road that is destined to be on the road for a while, and convert it to be able to use hydrogen and gasoline. Trucks of hydrogen liquid can be placed in gas stations before all the necessary pumps are up. The tankers could be suited to carefully filling a car tank. Time, technology, and discussion may dispel ungrounded fears. In that energy gap I spoke of over the next ten years, the hydrogen could be coming on board steadily. Harry thinks it could be and wants it to be done in five years. Well, I agree. But start thinking about taking the plunge into the simple principles, yet a whole new frontier. Perhaps a frontier is feared but it is also thrilling.

To be out aboard a boat and seeing the potent wind ships, in fresh ocean air, the thirty six turbines spinning rapidly in the strong wind. In playing with the photo I have, I did some zooming in one instance. I was almost as though on the ramp, seeing what looked like stainless steel poles, and turbines that I guess would have blown me away with their speed.

Turbines spin, wire coils in magnetic fields spin with the turbines, jostling electrons in the wire into a current back and a current forth, electricity that we call alternating current. Electricity is produced by spinning turbines. The wind is spinning them, not coal or methane or oil, as in the electric plant that heats steam to turn turbines. Hydrogen is not free—build the windmill though, and the wind will spin that turbine. It's a lot freer than gathering fossil fuels. Not all of the electrical energy is in the hydrogen, but what is in it, is a liquid fuel carrier of the energy left, once the hydrogen is released from water. It's a strong fuel. Our real expense would have been manufacturing the wind turbines, the wind ships. Engineered to be hurricane and earthquake proof by a Navy man, a Navy architect and engineer. Both talents went into the invention. Have a little trust. Those huge instruments will last a good long time, and every year we get out all we need, we are reaping the financial benefits of about one trillion dollars in consumer sales. I checked to see whether Harry was exaggerating—I was referred to a document with these figures up till 2001, by the DOE. In 2001, it was almost 700 billion in consumer sales.

That's an investment. You put in three trillion, not a wild estimate, and when in full operation, you have one trillion annually in sales. Read my letter to Warren Buffet in the chapter called, "Financing the Hydrogen Frontier," Chapter Eleven.

Well, back to making fossil fuel for transportation twice as available, and lowering the cost. Take the bus or train into New York, Boston, Washington, Baltimore, Philadelphia, etc.

I sure hope to see America all pitching in and conserving fuel in some way, to increase the supply by the abundance they didn't buy, so that it is less expensive for a person that can't afford it at all to get to the doctor. I want us thinking and planning, and I'm sure some communities might be, as my acquaintance and his church and political friends are doing in a Westchester town. My advice in this book is coming out as fast as I can write and polish it, but I guess mid-November will be the earliest. Once it's written, I guess I can start talking and think of public and media outlets for that. That should be by mid-October. Ah, Freedom!

This is the hardest book I ever wrote, for throughout the topic is so serious and the self-imposed time structure so much pressure.

I read a spoof of a Prius owner and a Hummer owner filling up at a gas station. They had some unpleasant repartees, and the Prius owner drove off, leaving the Hummer owner filling and filling and filling, thinking various thoughts, until he got scared he'd be late for work. I didn't appreciate the spoof, for it is no value to our unity to be expressing contempt. It is far more important at a time like this to think about saving lives instead of making spoofs about differences. We've got plenty dividing us, but hopefully this winter's awesome crisis will unite us, and help us heal wounds. Forget doctrine, fighting, exclusion, clanning together. We need national prayer, meditation, and unity. I just hope we don't have frozen babies and mothers in the dead of winter. I hope no one gets frozen. Maybe we could leave the shelters one or two days a week to be in our own home. Maybe we can afford enough fuel for a visit. We need some triumph songs—Roll on Columbia, Roll on.

I wanted to bring up a story that I heard at the eye doctor's office. I met a couple who had been in Poland, in the part of Poland that was ruled by the Russians, by an agreement with the Germans, before Hitler invaded Poland. Well, these two people said they were not afraid of dying, but both lost their whole families in Auschwitz. They themselves were rescued by the Russians and I think she said they were taken to Siberia. They worked in factories and had just a small loaf of bread to eat all day. I think that that was true for all the Russians, too.

I found the attitudes of this couple to be very positive. I had wanted to ask them if they had flashbacks. I asked, but the question was deflected by some other aspect of the conversation. I think they might not have been affected, too badly, in that respect, as they were such positive people. And it sounded like they were together, not separated, throughout. That can make a world of difference.

The man wanted to tell me a funny Jewish story—and I don't think it's apropos of us in America, but maybe it could be. If there is no policing, no rationing, no guide to what each person can do to save fuel—while also trying to think of making it to work, keeping their job, it can be very hard. If you can't line up a bus, go to the store with a neighbor or two. And cooperate, even if you are not one of the financially hard-pressed.

Anyway, to the story. A town leader announced that we'd love to have a great celebration in our town, but to do it, we need a lot of vodka. So we suggest that every family put a glass of vodka into a barrel, each day. Then at the end of the year, we would have plenty of vodka for the celebration. Well, there was one family who decided, let's put water in the barrel. It won't make a difference. No one will notice. Unfortunately, everyone decided it wouldn't make a difference, and put in water. There was no celebration. The story of the vodka made me think of the way the early church was organizing in the Book of Acts. Everyone that had land or wealth was selling it to share with those in need. There was only one holdout. A man and his wife held back some of the money. The both dropped dead. Do we have these kinds of Christians and other citizens in our country, the kind that don't hold out? Once the emergency is clear to us, I think many will see the ethics of helping one another and the need for national unity.

We may have trouble telling our own needs, from how to conserve, perhaps. Stronger leaders, such as church leaders and other community leaders, can help counsel people with their individual situation, and also get it in gear, if it's time to set up heat shelters. I think of having a day or two to go home, having left the thermostat say at 40 degrees, so the house could be warm on the weekend, and the water restarted, till Monday morning. A government advisor in Maine disagreed with this possibility in Maine. Once the water was turned off, she said, it couldn't be turned on until the shelter months were over. Her elderly citizens preferred to share the winter with friends or relatives, I presume that she meant that they'd be sharing the heating bill, too. I said, yes, I'd thought of that as house-pooling. She had not ruled out forming heat shelters, however, and spoke of having done so during an ice storm last year. The Red Cross got involved. When I spoke to the Red Cross, they said they were only available if the heat went down to 10 degrees or 20 degrees, and not on a steady basis. It seems to me that if the state or federal government gets involved in increasing the amount of heating funds available, they might be amplified in effect were people housed together in heat shelters. Maybe they could supplement house-poolers as well.

All people need to get versed in what our total emergency is about, our overall dilemma. The environmentalists, the workers, the businessmen, the religious leaders, the environmental engineers, the media, the colleges, the high schools, the junior high schools, the parents, and our local and Washington leaders.

God bless us all as we work at getting through the winter. The Christmas season is coming upon us, as is Chanukah. Ramadan was early this year. Maybe the winter will be mild, by surprise. That happened in 1979-80, at least in the Northeast.

I think next season will be easier, if we've put some new habits and ideas in place. It will take time to get organized this year. But next season, with or without the hurricanes, we'll have some practice. I don't expect the kind of storms that ravage the hurricane zone in the South, to be as potent in the Atlantic and Pacific. I don't think that the windships would be inadequate for the storms in those locations. They are built to withstand earthquakes and heavy storms, as they are ballasted by the heavy dome at their base, and if they dipped forward in the wind or waves, they'd pop up. Dr. Heronemus was a Naval engineer and architect.

We are going to manufacture cars by the millions, in the next year, I'm sure. We should get some prototypes of the wind ships up as soon as possible. Hydrogen for electricity plants and cars, perhaps for home and apartment building heating. Research or engineering can be done to maximize hydrogen burning equipment in many applications.

CHAPTER SEVEN:

SOME OF THE URGENT RESULTS OF GLOBAL WARMING
[IF YOU DIDN'T KNOW THE DETAILS, HERE IS WHY EVERYONE IS
TRYING TO AVOID IT]

SOME OF THE URGENT RESULTS OF GLOBAL WARMING
[IF YOU DIDN'T KNOW THE DETAILS, HERE'S WHY EVERYONE IS TRYING TO AVOID IT]

The date of my urgent new start, after writing to three politicians, one of whom was the President, was July 27, 2005. How could I quickly get out a book in time to be of help? By now in late September, people are conserving gasoline out of necessity and the demand is down some. That would be everyone's goal who is just making ends meet. Here, I hope the rich and the poor alike cooperate both with gasoline conservation and the heating methods this winter. And we certainly need to think about heat shelters for many people, at increased prices for heating oil, natural gas, electricity, and cord wood. [The heating oil reserves that the president plans to make available, in the Northeast, are only a 10 day supply, 10/6/05.]

Where does warming come into this discussion. I guess one could wonder if the ferocity of the hurricane season was not the result of global warming. Someone in the U.K. made that statement. We don't really know, but it could well be. [When I spoke to Ross Gelbspan, author of *Boiling Point,* [Basic Books, 2004] the other day, he was extremely busy. After doing an article with the *Boston Globe* on the issue of global warming in relation to the recent fearsome hurricanes, a relationship that a large number of scientists have been convinced about, he was requested to do about 45 interviews. *USA Today* permitted him to rebut an article in their paper which said some people were putting spin for their cause, on the hurricanes we just experienced. They said there is global warming, but no one should rush to say that the recent disasters were definitely a result of it. "Even the researchers who suggest there may be a link, caution against leaping to conclusions without lots more study." [*USA TODAY,* 9/09/O5] However, the National Hurricane Center notes that "the current cycle of more and deadlier storms could last 15 to 20 more years." *That* is bad news. It speaks of great human suffering, tremendous damage to coast and property, *and danger if we are dependent on Gulf oil, natural gas, and the 25% of our refining that goes on along the vulnerable Gulf Coast.* I read in one source, that we could have the refining done in other, poorer countries, where it would be cheaper, but how long does it take to build refinement enterprises? I believe it was also said that there is a general shortage of refining capacity in the world. We need refinement, even if only for five years or more, intensively, as we contemplate, hopefully, a shift to wind and solar, for hydrogen fuel.

I will review only some of the appalling findings about global warming that Ross Gelbspan incorporated into his book, *Boiling Point* [Basic Books, 2004.] I became aware of the book last spring, around the same time as some business men and national leaders, as well as faith-based groups spoke up about global warming. I'll quote the news clips that announced the faith-based action:

"EVANGELICAL LEADERS SWING INFLUENCE BEHIND EFFORT TO COMBAT GLOBAL WARMING", *The New York Times,* March 10, 2005.
"A core group of influential evangelical leaders has put its considerable political power behind a cause that has barely registered on the evangelical agenda: fighting global warming....The Rev. Rich Cizik, vice president of the National Association of Evangelicals and a significant voice in the debate said, "I don't think God is going to ask us how he created the earth, but he will ask us what we did with what he created."

And from *Christianity Today,* "Heat Stroke," editorial, October 2004:
"In the United States, Senators John McCain and Joseph Lieberman have introduced legislation that would set targets and create a trading system that would allow companies to reduce emissions in a way that is responsive to the market economy. This bill is an excellent starting point for whatever the Congress and administration will eventually develop...But time is important. With each passing year, we lose the ability to slow and minimize the effects of global warming." [These conclusions were very well worded—"slow and minimize the effects of global warming."]

Ross Gelbspan's recent article in USA TODAY, was entitled "Our Denial is at Category 5." "The question that most requires our courage to answer is: Which, in the long run is more dangerous—stronger hurricanes or our own Category 5 denial?"[9/25/05]

This is intended as a chapter to explain some to the initiate what global warming is doing, already. Let's leave the hurricanes as potentially dangerous for 15-20 years by the National Hurricane Center's predictions, for now. USA TODAY'S journalist did agree that we do have problems with global warming. It is caused by a "greenhouse effect." Maybe that phrase is a blank for some readers. What's a greenhouse? It is also called a "hot house" for people who know them at the level of farming. A greenhouse or hot house is a garden grown under a roof and walls of glass that let the weak winter sun in, but do not permit the heat to dissipate. In that way, tomatoes and other foods can be grown indoors in the winter. Now to "greenhouse gases", and global warming. The carbon dioxide is permitting the sun's radiation to enter the atmosphere but is acting like the glass in the greenhouse. It is not permitting the sun's radiation--induced heat--to escape as fast as it is entering. Hence a gradual increase in atmospheric temperature occurs and the oceans absorb some of that heat. The oceans absorb some of the carbon dioxide as well. As long as we go on burning fossil fuels, the rate of warming will increase, along with the negative results of that.

The most obvious direct effect of this is the warming of the glaciers, both in the mountains and in the polar regions of the Arctic and the Antarctic. The glaciers are fresh water—for the inhabitants in the region of glacial mountains this is causing them to eventually lose their crop and drinking water supply. Mt. Kilimanjaro is a case in point of a glacial mountain whose ice will be gone in twenty years, in the African country of Kenya.

At the poles, glaciers are melting at a rapid pace: As glaciers are fresh water, their melting changes the concentration of salt in the seawater. In   time, not ten years as we are inclined to react to the statement, but in 100-200 years as best as we can ascertain thus far*, the dilution of the salt water in Greenland and other northern regions, will stop the Gulf Stream from flowing. The Gulf Stream works its way from the area where the hurricanes just occurred, upward to the north, warming shores of North America and Northern Europe. When it gets cooled enough in the Northern regions, it sinks with its cold saltiness, and underneath the warm stream, it flows back to the South.

When the melting of the fresh water polar and sub polar ice has diluted the saltiness of the sea water enough, the heavy, cold salty water will no longer sink, and flow back to the south, to reheat and flow upward to the North, keeping North America and the Northern Europe and the United Kingdom warm. [There are places off the coast of Maine where the sea water is 70 degrees Fahrenheit as a result of this Gulf Stream warming.] The water won't be heavy and salty, as the ice melts from fresh water of Greenland, etc. and the Gulf Stream will cut off. North America and Europe will be in the deepfreeze. The U.K is at the same distance above the equator as Labrador, Canada. That's cold without the warmth of the Gulf Stream. Meantime, the Inuits who depend upon the ice being firm and the frozen ground [permafrost] being firm, the people we call Eskimos in North America, and their North of Europe relatives, are complaining about global warming, as the ice is giving way and their homes and other buildings collapse on the melted permafrost.

*When I asked Woods Hole Oceanographic Institution, in Woods Hole, Massachusetts, what the calculations were about the Gulf Stream cutting off, and that without much warning before it's suddenly more like Canada, here on the North East Coast, they referred me to a *Science* journal issue, 6/17/05, written by Ruth Curry and Cecilie Mauritzen. It confirmed that the cause of the problem would be glacial melting—Greenland is a glacier! At the observed rate, it would take 100-200 years,... but uncertainties remain. That's not far off, but when I read about it, I was feeling it was ten years off. One thing that concerned me was that going by change in weather alone, we might have little warning. But, in contrast, the Ruth Curry and C. Mauritzen article ended as follows:

"At the observed rate, it would take about a century to accumulate enough fresh water (e.g. 9000 km cubed) to substantially affect the ocean exchanges across the Greenland-Scotland Ridge, and nearly two centuries of continuous dilution to stop them. In this contest, abrupt changes in ocean circulation do not appear imminent.

"Uncertainties remain in assessing the possibility of such disruptions. A weakened Atlantic Meridional Overturning Circulation in the 21st century is a feature of numerous climate simulations of greenhouse warming. The cause is similar in all the models: glacial melting, enhanced precipitation and continental runoff, which are projected to increase freshwater input to the Arctic and sub-Arctic seas. Pooling and sudden release of glacial meltwater, disintegration of shelf ice followed by a surge in glacier movement, and lubrication of the glacier base by increased melting are all possible mechanisms that could inject large amounts of fresh water from Greenland's ice sheet into the upper layers of the Nordic Seas. The possibility of such events precludes ruling out a substantial slowing or shutdown of the overflows as a result of green house warming."

In *Boiling Point,* [p. 91], Ross Gelbspan makes the following quotation, "The very recent freshening signal in the North Atlantic is arguably the biggest and most dramatic change in ocean property that has ever been measured in the global ocean. Already, surface waters in the Greenland Sea are sinking at rate 20% lower than in the 1970's," wrote Robert Gagosian, head of the Woods Hole Oceanographic Institution, in 2003. "At what percent will the Ocean Conveyor stop? 25 percent? 40 percent? 60 percent?" [He goes on to say that rather than being like a dimmer-switch it would probably be like a light switch: on then off. The women in the above article, two years later, Ruth Curry from the same oceanographic institute, allow for the possibility of slowing or shutdown, suggesting a *possible* gradual effect to me.] When I spoke to Mr. Gelbspan this afternoon, he said that he agreed with me that maybe it is not yet possible to be sure of what trend things will take, decades or a century to two.

If the change were to be a sudden switch off, without gradual temperature change, the oceanographic and satellite studies of the observed rate of melting might give us warning as to whether to migrate, or give us a chance to figure out how to adjust. But please! We need a fuel that we can depend upon to migrate or to adjust. With fuel, we could travel by snowmobile as the modern Inuits do. We could make phytofarms, which are farming indoors, with controlled conditions, enabling us to grow 100 acres of some plants in a one acre building, without soil, using less water, and getting faster growth than in normal farmland. We need clean, renewable hydrogen, if we plan to keep a lot of people alive. But our complex urban societies will not be viable, with such a change. If it's going to be more and more coal, with more and more pollution, or shale oil, tar oil, etc., that will put us beyond the boiling point. We are already just hoping for the capacity to manage ourselves. As *Christianity Today* put it in the editorial, we need to "slow and minimize the effects of global warming."

Global warming has already happened. We have to try to live with what has developed already, hoping that prompt correction of our fuel use here and around the world, will make things easier for us. The Arctic ice was 20% less this September, than it usually is at this time of year.

Can we put windships in the Atlantic, I wondered. It was one of the things that prompted me to call Woods Hole. The windships might eventually be endangered by icebergs, if the Gulf Stream switches off. The windships could be moved South, if necessary. They are tethered, but not attached to the ocean floor. By that time, we'll have some idea of how warm the Gulf of Mexico is. How will it change with deepfreeze to the North? This gets fairly complicated. And this was the easy part. There are many things about global warming that effect the weather. Tornadoes, hurricanes, droughts, la nina, el nino.

"Unintentionally, we have set in motion massive systems of the planet (with huge amount of inertia) that have kept it relatively hospitable to civilization for the last 10,000 years. With our burning of coal and oil, we have heated the deep oceans. We have reversed the carbon cycle by more than 400,000 years. We have loosed a wave of violent and chaotic weather. We have altered the timing of the seasons. We are living on an increasingly precarious margin of stability." [Ibid., p.4]

"Three years ago, a team of researchers reported in the journal *Nature* that unless the world is getting half its energy from noncarbon sources by 2018, we will see an inevitable doubling—and possible tripling—of atmospheric carbon levels later in this century. In 2002, a follow-up study by many of the same researchers, published in the journal, *Science,* called for a Manhattan-type crash project to develop renewable energy sources—wind, solar, and hydrogen fuel." [Ibid., p. 4]

There is dovetailing between the sudden need to conserve fuel to deal with the rising world demand and our own domestic demand, in the face of prices that are threatening to put us in progressive economic and human distress—and the urgent need to get clean alternative energy to replace declining fossil fuel sources. To stay warm and mobile we need non-carbon sources of energy. The "dirty" alternatives take time to develop, too. We need to fill an energy gap in short order. The possible exception is, of course, coal. But now that it's been acknowledged that carbon dioxide is a greenhouse gas, along with methane and sulfur in the air, the energy companies promise to try to find ways to sequester CO2 now, and to get us hydrogen from coal, and sequester the carbon dioxide. I am amazed that so many people in the search for alternative fuel, go along with the idea of carbon sequestration.

Ross Gelbspan, in *Boiling Point* [Basic Books, 2004] points out that the Apollo Project that appears progressive has several fallacies. The first have to do with not taking into consideration animosity of the oil-producing countries if we suddenly stop our use of their products, without helping them to also be developers of renewable energy sources, when poverty is one of the main causes of the anti-U.S terrorism. In addition the Apollo Project has chosen to promote the "single largest contributor to global warming—our burning of coal. Coal emits about 30 percent more carbon per unit of energy than oil and about twice as much as natural gas. There is no quick way to address the climate crisis without quickly eliminating coal from the world's energy diet...(It seems that compromise could have been avoided by the inclusion in the Apollo Project of a fund to retrain or buy out the nation's 50,000 coal miners—perhaps by siting some wind-manufacturing facilities in coal-rich areas.
...Because of its politically motivated inclusion of coal, the Apollo Project also regrettably calls for huge investments in carbon sequestration. Under this strategy, carbon dioxide would be captured from coal-burning power plants and piped into burial areas deep inside mountains or mine shafts. It is an extremely risky and unproved method of reducing atmospheric CO2. More to the point, it is extraordinarily wasteful...", non-effective and at costs that are economic lunacy.

"Unfortunately, deep-mine burial of carbon dioxide is a very uncertain and expensive technology that could lead to massive releases of CO2 in the future. It could also contaminate aquifers [water sites held in rock or soil layers], increase the likelihood of earthquakes, and damage biological communities within storage sites, according to the Union of Concerned Scientists." It should not be researched at the expense of alternatives and it would probably double the cost of electricity. [Gelbspan, pp. 164-169.]

Gelbspan advocates international governments "to regulate the transportation and energy industries so they can make the transition into a new regime in lockstep—without any company sacrificing its competitive standing within its own industry." [Ibid. p.168]

He sees the climate crisis offering us an extraordinary and revolutionary opportunity to have a global crash program to rewire the planet and reframe the energy infrastructure to contain seeds for a wealthier, more democratic, and ultimately more peaceful world.

.   .   .

There are other issues which I will address now. The first is that there has long been a debate about whether it is human activity that has caused global warming. Ross Gelbspan played a large part in revealing that the denial of our place in global warming was paid for and fueled by the energy industry itself. Our role in global warming is a fact determined. Countering it is a propaganda campaign from the oil and coal industry, especially Exxon-Mobil and Peabody. They have been instructing people with no prior education on the matter around the country that they should be skeptical about our role in global warming. The people writing these opinions have no respect in the scientific community. Yet their campaign had a marked misleading effect. The number of people who thought global warming was a very serious problem in a *Newsweek* poll in 1991, was at 35%. By 1996, when the scientific community's gleanings were far more robust, the success of the propaganda campaign had reduced that number to 22%.

Recently, having lost ground publicly, in this debate, Exxon-Mobile has reduced its objections to saying things like global warming is good for us. All of this was exposed in Gelbspan's previous book, *The Heat is On* and is summarized in the present book. He goes on to say that Bush's participation in gutting efforts to participate in the issues of global warming is corruption. Some justify it as politically conservative, "the withdrawal of onerous regulations, a belief in the unqualified efficacy of free markets, and the tepid appeal to corporate voluntarism....In fact, the Bush climate policies have nothing to do with political conservatism. Rather, they represent corruption disguised as conservatism."[pp. 42-43, Ibid.]

Gelbspan points out that there are two other sectors guilty in the misrepresentation. One is the journalists, who gave equal time, rather than finding a way to communicate with the scientists whose work demonstrates human induced global warming, in such a way that they could understand the issues. And the environmentalists were compliant, too, in unwitting compromises.

What we need is a 70% reduction in the burning of fossil fuels, if we are to keep global warming under future control, according to Gelbspan. The oil and coal lobby know that this would put an end to their industries. Hence the intense fight toward deception.

The chairman of Peabody coal had donated $250,000 to the Republican National Committee for the 2000 presidential race. In payback, Cheney declared that coal had been neglected as the "most plentiful source of affordable energy" in the United States, adding that people who sought to phase out its use "deny reality". "Conservation may be a sign of personal virtue, but it is not a sufficient basis for a sound, comprehensive energy policy." [Ibid., p. 45]

Coal-mining is the most dangerous field in energy employment—there have been more deaths by far associated than with even nuclear accidents. It is not necessary to use coal, and it is not necessary for the young people of West Virginia to continue in the coal mining tradition. They could be trained to work in clean energy programs. Yet West Virginian was a swing state that put Bush in office in 2000.

At what point is it that the energy companies realize that they too will be destroyed by global warming, by the fact that they are getting behind in their ability to satisfy the energy demands of the country and the world? Hanging on is human nature, I suppose, but at what point can they make a transition to alternative fuels? Gelbspan and Braun both seem to want the current energy companies to be the energy companies of alternatives. I question that. The tendency to start to dupe the public by saying that they can get hydrogen from coal and oil, leaving out that this is equally polluting, does not speak well of their preparedness to see clearly and move forward with wind, solar, and hydrogen, 100%. I say 100%, not 70%. What do we need the other 30% for, if we can avoid it?

On November 24, 2005, it was published that a letter of plea came from a group of U.S. senators, addressed to nine big oil companies. "With huge increases in winter heating bills expected, the letter read, we want you to donate some of your record profits to help low-income people cover those costs." The only positive response came from Citgo, from Chavez, President of Venezuela. It is seen alternatively as an altruistic move and a political move. But in helping to pay the average U.S. household bill that's increased 27% over last winter, it makes a difference in Boston and the Bronx where the discounted oil was headed to go. Eleven Senate Democrats turned to oil companies after Congress did not add funding for a low-income energy assistance program. [from a web news site, notjustanotherclone@yahoo. com.]

How do these oil companies benefit democracy and national and international survival? They are the epitome of social irresponsibility when they encourage a self-interest that is profit-seeking only, and not in any way showing benefits to the needs of the people, and especially the needs of the planet's stability. They and their children will be victims of the overall tendency to destruction of civilization as we know it. We, as Americans, must look to our own capacity for unity, around what is this winter's hardship, and probably next winter's, too. I don't think that Mr. Gelbspan's hope to rewire the developing countries, first, with clean alternative energy economies will happen in that order. We are holding the world back from Kyoto type improvements. We are using 25% of the energy. We are making 25% of the world's emissions, a lot on the road. [20% of U.S. emissions] And I was shocked to find out that the ostensibly progressive Apollo Project plays right into the hands of the coal industry.

We are seeing a growing cost to mobility and electricity and heating. If we are rendered helpless, we will not be able to help the underdeveloped countries towards the natural growth of prosperity they would have with energy alternatives, and all the jobs that would create. It is not a mistake to put up our own global warming reduction equipment for alternative fuel promptly. Out of that same growth of good will amongst ourselves, the steps of taking into consideration the needs of the oil producing countries and the underdeveloped countries can be strengthened. Our leaders in Washington have duped us to break down our part in democracy, giving the average person the burden of taxation, and not the wealthy. This is not Christian. This is not democratic. This is not Judaic nor Islamic. This is not humanitarian. This is not smart. This is corporate protection, welfare where there is no need. Destined for destruction, for we have only borrowed income to pay for a war, and little left for benefits for the people. Stranding the Katrina Victims in the hotel at Kennedy, and leaving the heating casualties to no one, so far, except the more foresighted local communities. Maine is raising money from some in Maine to pay the heating bills of others in Maine. Charity, after taxes were paid. There are many charitable people in Maine. But what about the extra cost they have for their own large homes, if that's what they are living in?

.   .   .

I could quote more from Mr. Gelbspan, more about global warming, that is not subtle. In fact, it's so dangerous to civilization that one does not want to know about it. It's easy to take a position of denial. That was going on over the fuel crisis, too. Many have been sleep-walking. But now, in the month of November, the human factor is on the air. Earlier articles often mentioned the threat to economic growth, with a fuel shortage. People clearly didn't appreciate the threat if they were buying an SUV in July. To manufacture an SUV with low mileage is insane, also. The car companies were just as aware of climbing prices of oil, as I was. Why do we need the government to enforce higher gas mileage? It doesn't cost Detroit a lot of money to increase gas mileage.

I'm going to review the synopsis of events that Gelbspan goes over for the year 2003. I'll start by saying something that has yet to be said here.

"Nor was it only the planet's physical systems that felt the threat of escalating climate change. Many financial institutions also began to feel the heat in late 2003.

"Pension fund managers, bankers, and Wall Street advisers—representing more than $1 trillion in assets—issued a 'call to action' in November 2003 about impending climate-driven upheavals in the world's financial markets. At the meeting, which was sponsored by the United Nations, the treasurer of the state of California, Philip Angelides, declared: 'In global warming, we are facing an enormous risk to the U.S. economy and to retirement funds—that Wall Street has so far chosen to ignore." They tried to persuade businesses to address problems that would cause global warming. [pp.9-10, Ibid.]

Global warming, thus not only threatens the planet's climatic stability, but along with it, any business endeavor that gets in the way. It's a global catastrophe that is more than one would like to contemplate. In the fall of 2003, several developments were examined:

"1) The entire ecosystem of the North Sea was found to be in a state of collapse because of rising water temperatures. [Their normal fish species and supply for fishing economy and food were ruined, and replaced with fish that were more southern in type.]

"2) For the first time in recorded history, the world consumed more grain than it produced for *four years in a row*. The reason: rising temperatures and falling water tables—both consequences of global climate change.

"3) The German government declared that the goals of the Kyoto Protocol need to be increased by a factor of four to avoid 'catastrophic' changes. Otherwise, the climate will change at a rate not seen in the last million years.

"4) The most highly publicized impact of global warming in 2003 involved a succession of headlines from Europe about an extraordinary summertime heat wave. Scientists attributed the unusually high mortality rates not to the fact that the August temperatures were so much higher than before...The findings took on an especially grisly reality. The lingering nighttime warmth in Europe that summer deprived overheated Europeans of the normal relief from blistering daytime temperatures...When that brutal summer finally subsided, it left more than 35,000 people dead.

"5) The following month...the biggest ice sheet in the Arctic—150 miles square in area—collapsed from warming surface waters in September, 2003....A massive freshwater lake long held back by the ice also drained away." [Ibid., pp.7-9]

6) The oceans have become acidified over the last 100 years because of the fallout of emissions from coal and oil burning.

"7) By the fall of 2003, an eighteen-month drought in Australia had cut farm incomes in half—and left many scientists speculating that the prolonged drought may have become a permanent condition in one of the country's richest food-growing areas." [Ibid., p.9]

.　　.　　.

The Christmas songs are coming over the air and the television. The shopping reports for the outlook on consumer spending have started. I had to rewrite this chapter, primarily because the first time I was avoiding some of the ghastliness. It is very depressing. I remember that Ross Gelbspan said he was often overcome with hopelessness, writing the book. Someone said to him [a paraphrase], if we lose it, there will be some survivors and hopefully they will carry the truth with them, the right attitudes. Ross Gelbspan has the attitude that we must try, because we don't know everything, and there may be some good outcomes. Again, a paraphrase. He also senses what I do, that in trying to take the first steps to transform the world's energy resources, we shall begin to have a change in the way we think, that is more realistic with other problems we bump up against, resulting in more world peace and prosperity, and problem solving.

I was in Church this afternoon. The sermon was about signs of the apocalypse, the coming of Christ. They were not expressed exactly as this chapter is expressed. I gave lipservice to hope from God the last time I wrote the chapter. The priest said, It is time for an awakening. That's what Ross is saying in his way, and I am thinking in my way. We both marveled at the transformation of the Apartheid in a miraculously short time. We both feel that there are growing swells of concern about these issues throughout the world.

I think that there is a need for awakening in the United States. I plead for it in how we face the fuel shortages, in asking every American to participate in conservation. In asking the Mercedes driver, to use the bus or Metronorth. In stating that the heating shelters, if ample enough, could house the Mercedes driver, too. The heating fuel we don't burn would start a reserve. The methane that didn't get used up at the house while in the heating shelter, to generate electricity, would also be reserved.

We've got class differences, racial differences, religious differences, political loyalties that run deep emotionally, more about abortion from some citizens, and about the right to choose from others. We have sheer sarcasm and smearing in so many of our elections. But can we somehow look at the new challenges we are just noticing, because it's getting cold, or because we had to use our credit card to buy gasoline to get to work?

We've got some orderly steps to take. Making sure that every American has sufficient heat this winter—that is something that communities across the country are beginning to think about, or maybe don't know how to think about. I'm not sure where we'll be going with it this winter in New York, but I did find out a bit of information calling the heat and hot water line. The landlord does not have to heat after 10 P.M. unless it is below 40 degrees. The inside temperature between 10 P.M. and 6 A.M. does not have to exceed 55 degrees. Daytime temperature is set at 68 degrees. Is this new? Fifty five degrees is pretty cool. But at 10 P.M., it was 68 degrees, so it takes awhile for the temperature to drop. I have to investigate whether this is new. Usually, after some latish hour at night, I don't think we had heat, many times, in the past. It was warm enough in bed if properly clothed and covered. I couldn't inquire over the holiday weekend. It was stated that these temperature requirements have been in effect for years.

So trying to share the burdens of heat and mobility are this winter's objective, from my point of view. Trying to look at these as regular companions to be concerned about in the foreseeable future, so why fight about everything else we can think of? There is lots.

And we have a war going on. Feelings about this run very deep as well. But the weapon of mass destruction, not having fuel enough, is going to threaten more people than the war, over time. So without focusing on the Global Warming gloom, though we have to take the same steps I've been recommending regarding that, we have our immediate challenges. I think that any good cooperation that develops between people over fuel and electricity can give us access to the closeness and comraderie of each other. That can grow into action about alternative fuel and start the ball rolling towards cutting the CO2 down 70-100%.

I came away from church with more straightforwardness. There does need to be a spiritual awakening in this country. Someone who appeared quite ordinary announced on television that she spent 7 thousand dollars on Christmas presents this year. I'm sure she doesn't know about the heating problems in Maine or Milwaukee or Oklahoma—lets add Canada and Scotland and the U.K. and China. She's not trained to think about things like that. Someone said recently, I feel sorry for my poor mother in Long Island. Apparently, that mother has heating problems that we don't discern here in New York yet.

The American citizens responding to the fuel crisis and to the warnings of warming, uniting to save ourselves, not unmindful that there is a worldwide shortage of oil can start the ball rolling for the most positive future possible. Not unmindful of the Chinese or the Arabs. Hopefully, before Congress ever passes a Tobin Tax to discourage meaningless investment in currency speculation and giving three hundred billion dollars to underdeveloped countries for energy transformation, we can appeal to our banks and investors to ask them to invest in a real economy, the infrastructure for hydrogen from wind, from sewage and animal waste, from landfills and from garbage. And soon solar will be competitive with the first two. Three hundred billion from energy sales profit would achieve the same goal, if we gave it away. It is so evident that global warming will destroy the economies of the world, and it is so evident that high fuel bills will as well, that I should think it not that hard to shift investment for our major investment players. We need to be sure of what infrastructure we'll invest in. But we cannot look at that forever. We need to make some decisions, test them, and go with them. Certainly wind for hydrogen—the equipment I suggest needs testing, but very quickly, not later. The land turbines would do it, but so much more expensively, taking a great deal of land.

I genuinely believe that Exxon-Mobile and Peabody Coal are going to suffer greatly, with the rest of us, by the disorganization of structures if we don't plan wisely to get through the emergencies ahead starting this year. Five years earlier than Niall Ferguson's article predicted in 2001, the combination of prices is already out of reach.

Since 9/11, New York is a changed place. Hardship can bring great blessings. I watched the change burst open like a flower. Now it is a pleasure to go out and about. It is more like Europe now and more like the friendly cities and small towns of America with which I am familiar. Jersey City was warm when I was a child, and I loved New England. Philadelphia and Delaware and Maryland and Washington, D.C. were warm places. I guess that so was Durham, N.C. though I was there only a day, on the Duke University campus. The people I met in Tucson and Phoenix were friendly. San Francisco was warm. Everywhere I call, people are spontaneous.

What does this have to do with global warming? To proceed in the darkness of whether it is too late to correct our situation, if done with courage, with faith that we'll be trying together, using our best methods, not our worst, is a spiritual as well as practical exercise. I think that there may be some unforeseen grace, if we do this in humility. I know that nothing can separate us from the love of Christ, of God. He'll lift us from the disaster, by changing us and problems in ways unknown. But I do not believe that we are to stand back passively, doing nothing about our patterns of gaining energy, grinding what's left of Creation into the mud by lack of attention to facts and details. It seems to me that the Lord's Prayer would require keeping the Earth alive for His Kingdom to come on Earth as it is in Heaven.

We have no guarantees, but as we work together in good faith, we will have each other. I believe what we truly start, the enemy at our backs [fuel shortage, global warming, etc.] and the Red Sea before us, if we step into the waters, the Sea may part miraculously for us and our salvation.

CHAPTER EIGHT:

NOT TO BE NEGLECTED DISCUSSION OF PHOTOVOLTAICS
AND SOME OTHER SOLAR TECHNOLOGIES

NOT TO BE NEGLECTED DISCUSSION OF PHOTOVOLTAICS
AND SOME OTHER SOLAR TECHNOLOGY

I refer you to Braun's general statement that he made about energy economics, at the beginning of his chapter on "Renewable Energy Technologies." I suggest that you read that chapter in his book, previously cited, for a sense of how we are generally misled about the cost of renewable technologies. We think we will be paying three to four times as much, and though we researched alternatives for many decades, we always went the "market way", misled by the cost calculations for the "market way." Braun's book is replete with intelligent and well analyzed remarks about alternative technologies. He is under-read and highly informed; I would not recommend that anyone who comments as an energy expert do so without a thorough knowledge of his writing. You can see a picture of the cover and perhaps read some there or order there, at [phoenixproject.net]. Happily I can say, that I read and reread that book, and often use it as a reference.

As Harry said at the end of the passage I've just referred to, photovoltaics are the most widely known and mentioned in the media, and given most attention by the government, but are the most expensive of the alternatives that he investigates, and for which research and development has been done. Yet many Americans, Europeans, and maybe others are increasingly installing them in their homes or even towns. There is some cost reduction in mass production. However, they had been only approximately 10% efficient. Now they are more like 15%-20% more efficient.

PV cells are made with silicon, an electric conductor and insulator, then coated with a positive electrical layer, such as boron, according to Braun. The silicon is negatively charged and the boron positively charged. When a unit of sunlight, a photon, strikes a silicon electron, the electron passes to the positive part of the junction, and a direct current flows of electricity. The efficiency has made them inadequate as a good source of electricity, compared to other solar and to wind technologies. Their cost has gone down progressively, but has a long way to go to compete with other alternative solar and wind technologies. As of the publication of his book, in 2000, solar engine systems, similar in appearance to radar or satellite dishes, have held the world's efficiency record, since 1984, for converting "solar energy into grid-quality electricity."[See pp. 164-186, ibid.] Now, photvoltaic troughs lead for grid quality electric plants. [See Notes on ch. 8]

In a recent copy of the *International Journal of Hydrogen Energy,* two articles were devoted to some new solar hydrogen methods. In April of 2005, the article, "Solar-hydrogen: Environmentally safe fuel for the future", by J.

Nowotny, C.C. Sorell, L.R. Sheppard, and T. Bak, from the Centre for Material Research in Energy Conversion, School of Material Science and Engineering University of New South Wales, Sydney Australia, we find in the Abstract:

"There is a growing awareness that hydrogen is the fuel of the future. While hydrogen can be generated using different technologies, only some of them are environmentally friendly. It is argued that hydrogen generated from water using solar energy, solar-hydrogen, is a leading candidate for a renewable and environmentally safe energy carrier due to the following reasons:

Solar-hydrogen technology is relatively simple, and therefore, the cost of such a fuel is expected to be substantially less than that of the present price of gasoline.

The only raw material for the production of solar-hydrogen is water, which is a renewable resource.

Large areas of the globe have ready access to solar energy which is the only required energy source for solar-hydrogen generation."

The abstract goes on to say that the development of solar-hydrogen requires new photo-electrodes that are likely to be made of inexpensive polycrystalline material—determined by progress in the science and engineering of materials interfaces. "The present paper briefly outlines the main challenges in the development of materials for solar-hydrogen."

I want to note here, the information that has been released by NanoSolar, Inc. [nanosolar.com]. It may be *critical* information:

"Nanosolar, Inc., is focused on making solar electricity ubiquitous by delivering the world's most cost-efficient solar panels with process technology of unprecedented production volume scalability.

By making it possible to simply print solar cells that can deliver as much energy and lifetime as conventional silicon cells, Nanosolar has developed the capability to dramatically multiply the process through output possible in the production of solar panels—and achieve unprecedented cost and production volume scalability advantages.

Nanosolar's products are designed to streamline integration, distribution, and installation. Depending on a customer's system configuration, this can realize significant further total system cost savings. The result is a fundamentally better way of powering the world."

I understand that these are plastic polycrystals, coated with a semiconductor, and hopefully made at 1% of the cost of PV's and made 5x's as efficient as PV's. That would be 50% efficient. The source of the hopeful prediction in this paragraph is of unknown [to me] authority to say so. However, it stands to reason that more is better, in smaller dimensions, so both efficiency and cost, because of simple printing out, are greatly improved.

I reacted to the article in the above journal by thinking, I hope they find this cheaper way, but why do they neglect wind, entirely. Yet were wind and solar both unleashed, we would have hydrogen in great magnitudes of current energy supply. But we've got to plan, nonetheless, to provide energy for the 2 billion who have none, and for the people in China and other locations, no doubt, that have no heat in the frigid winter. Public transportation should precede cars for the more affluent in a given developing nation. We are in a position to shift our example some, on many scores.

Some people are advocating bicycles, and many are using them now. Motor scooters and mopeds seem to appeal to men as an alternative. I rather like them myself. They could be built with weather protection. Or one could wear a suit that would hold up in very cold climates, as weather protection. The Chinese have had to fight the government to retain their electric bikes, that have become very popular because of the pollution. China wanted to outlaw them on the roads, but they gave in to popular pressure.

Someone said in writing, the fuel cell could be important to the affluent and leave out the common man from being able to drive a car, with a shortage of energy.

[I was very struck, in reading several copies of the fore-named journal, that wind as a source of electrolysis of water has *never been mentioned, as yet.* As there is diversity in the energy and ecology fields in general, one hand not knowing what the other is doing, or at least not understanding what the other is doing, there are diverse approaches to the articles and studies published in this journal. However, this issue does contain an article that is very inclusive of issues that are involved, in trying to start a hydrogen fuel supply that is renewable.]

"Hydrogen as fuel: a critical technology?" by S.Z. Baykara, Chemical Engineering Department, Yildiz Technical University

I will note just two issues leading to critical questions:
"1.6. Government policies—Governments will have to balance policies involving promotion of hydrogen energy system, discouragement of fossil fuel utilization, and increased energy efficiency and conservation (these activities might be less challenging than starting life on Mars perhaps...)."

"A critical result of not being able to implement hydrogen technologies soon enough could be a very expensive delay in establishing sustainable development."

.   .   .

I must mention that the state of Hawaii has a website that is interesting. Their opening statement points out that they are very isolated, cannot turn to another state for oil, must import 90% of their oil, and thus, they are very interested in alternative fuels. One source is geothermic, another is biomass, and a third for them to consider is OTEC ships. Ocean Thermal Energy Conversion ships, that is. Braun spells *this very surprising solar technique out in detail in his book.* In essence, the world's solar heated pools can be harnessed by a technique one would hardly think of, if not an engineer by aptitude or training. By drawing up the cold water, 2000 ft. below the 80 degree water, one can boil ammonia at a low boiling point of 80 degrees *when the 40 degree water warms to 80 degrees.* This results in the capacity to turn the turbines with the boiling ammonia gas, creating electricity. Hawaii does not mention making hydrogen with the electricity, but they perfectly well could. The results of this process, if done with water that is evacuated at the 80 degree point, causes flash vaporization of the water, producing a huge amount of desalinated water for drinking or irrigation. It operates steadily 24 hours a day, at 1% efficiency, but Harry Braun believes that is enough to take care of the world's current energy needs, in itself. I often wonder whether the OTEC desalination and hydrogen production in the Gulf of Guinea in Africa couldn't help the drought in Niger and Borkino Faso, not far from the Gulf of Guinea. This means of producing fresh water is much cheaper than the method currently used, i.e. osmosis. The water and the shellfish that would be fed by plankton drawn up from the cold water would more than pay for the infrastructure. Well, Hawaii is interested in OTEC again, after many years. I think, since the 1970's. OTEC is detailed again in the Braun's book. Dr. W.E. Heronemus was researching OTEC in the 70's. He is the naval architect and engineer who invented the multiturbine windship illustrated on the cover of my book.

For those who understand how it applies, the OTEC heat difference, allows for the Second Law of Thermodynamics to operate. I am not at all clear about this—it has to do with temperature difference. But.....I wonder where in this book I lost most people. It has lots of information. By chatting, my preaching about conservation, my updating the costs of fuel, I perhaps give the average person a break. But these things that are technical are being ignored, and people are trying to reinvent the wheel. I suppose, I did some too. I got most involved with the idea of turbines that could turn with a small flood control dam in the areas where they were built to protect against the hurricanes in PA and MA, and maybe elsewhere during the Eisenhower administration. It so happens that the dam on the river next to our home in Massachusetts has elevators for the salmon that were stocked in the river, not too long ago. Bass were stocked there too. We just had brook trout and suckers when I was a child. The trout were stocked for every fishing season, too. The river was polluted by sewage and green waste from the factories down stream. Near the house, it was crystal clear. Now technically, that was a river, but an adult could easily throw a stone across the river.

.   .   .

Let's get back to Hawaii. They talk of burning biomass for energy, every kind of sugar and molasses and all the things that are either in lesser amounts as residues or larger amounts that may be interfering with the raising of necessary food crops.   Are they farming like everyone else, and using fossil fuels? Maybe there is more manual labor as in Brazil. Other biomass that they use is methane from landfills—*a very good choice, since methane from landfills is much worse than CO2 in the atmosphere.*

Their last significant biomass is wood. This is also polluting, Just think of the smell in the air of a wooden house on fire. But as Dr. Pimentel said, we may need to use everything we have for the while! Wood would qualify, in wood chips, as energy positive, as would be some other crops that Hawaii said could interfere with food crops. Wood can safely be reaped at a rate that was realistic for new trees to mature. Wood smoke can be scrubbed and maybe should be. It doesn't make sense to burn the crops from land that is needed for food; plant residues in the field need to be used as compost. There is perhaps a percentage of the scraps that could be spared for the thermal heat of biomass, but this should be distinguished from *biofuels, that we have seen in a previous chapter are energy negative and pollute with the fossil fuels used to cultivate and irrigate them, and distill them. Biomass refers to plant material used for gaining heat, and was never distilled. It is not the same as the energy intensively distilled biofuels that required so much fossil fuel for their preparation.*

Biomass is plant residues, plant crops, wood, landfill methane. *Biofuels* are alcohols made from various plant crops. They are distilled. Switchgrass pellets are compressed plant. That is biomass and unlike biofuel it has not been distilled into alcohol. The biomass is used for cooking, heat, or other ends. I refer you to an article that I received today from Dr. David Pimentel, who discussed biomass as an alternative renewable energy. The article is a review of renewable methods as he understood them in 2002. *Renewable Energy: Current and Potential Issues, December 2002,* in the journal *BioScience, p.1111.* If you email him at [dp18@cornel.edu] you can request a copy.

In this same paper, he mentions parabolic trough photovoltaics that heat some substance—perhaps the oil mentioned in BMW's discussion of what they thought was the most up to date method of solar energy production, in the Mojave Desert. Parabolic means trough like constructions of the photovoltaic cells.

.   .   .

It would be wrong to discuss solar energy without mentioning passive solar energy—both retaining heat and promoting cooling can be accomplished by proper buildings. It is harder to achieve with already thoughtless homes with regard to this. But in new buildings, it can be planned in. That's one reason we should probably stay where we live, and not rush to buy a home, until some of our energy issues have been sorted. Where is it best to live and what kind of housing would be best to heat in winter and cool in summer? I'll give just two examples of homes that are self-sufficient. Besides the wonderful worship services we had at the Cathedral of St. John the Divine (Episcopal) in New York City, in 1978 I got a lesson in energy. John and Mary Todd visited to tell about a home they devised in Woods Hole, MA that was basically a greenhouse, in that it was all glass, with a windmill for electricity, not a big one, I don't think. It was warm for the inhabitants, warm for a full vegetable garden, and contained a substantial fish tank for edible food. Woods Hole is pretty cold and windy in winter, at the south side of lower Cape Cod. Next they built a similar house, but with some architectural wood along the edge of the roof. This was on Prince Edward Island, just north of Nova Scotia. That is even colder to be sure. This house was also self-sufficient.

A knowledgeable friend told me, also in 1978 or early 1979, that a home in New Jersey was built with a black wall to absorb the heat.

Then there are the more familiar things, like well-sealed windows, in our present housing.

NOTES ON CHAPTER EIGHT

The reporter who did a critique of my book, prior to publication, brought it to my attention that the reason that PV's have improved from 10% efficiency, to 15% or 20% efficiency is that there has been development of "thin film cells", made not of silicon, but of copper indium selenide, CIS, a much more efficient collector of light than silicon. One company whose methods stand out as the gold standard is Heliovolt.

Summarizing some material from a press release from Heliovolt and some information from their website, dated 11/10/05, I learned that worldwide solar energy revenue is expected to grow from 4.7 billion dollars last year, to 30.8 billion in 2013. The price of solar panel dropped from $100/watt, 30 years ago, to less than $3 dollars today, and Heliovolt promises to get the rate down to under $1.00.

The vast majority of PV's today are manufactured with silicon. Copper indium selenide collects light far more efficiently. Most thin film cells were made of plastic in the past and did not hold up in the sunlight. This is not a problem with CIS. It can be incorporated into the building materials, not added on by bolting or retrofitting. It can be in the roofing, the glazing, and the curtain walls. There is speed in the manufacturing process which also lowers the cost.

Shell Solar Industries, a chief competitor of Heliovolt, recently announced plans to boost production of copper indium cells.

The goal of the Heliovolt company is to build cells into the construction material of a conventional new building, and not have it retrofitted after construction. This promises an excellent future for self-sufficient buildings, requiring no electricity from the common pool, and as such helping with the alternatives to fossil fuel electricity. It has great promise. It is still developing, and doesn't promise to cut our crisis immediately, but does promise a clean future of solar power.

I don't know whether the technology could be applied to generation of electricity in solar hotspots in the world, to decentralize hydrogen production .But wind or solar, there is very likely not to be a monopoly in any one country or several on energy independence. All countries have wind, some countries have ocean wind, and some countries have very intense solar heat reaching the ground. The solar future that Heliovolt promises, does not depend on intense sunlight. It makes buildings independent in temperate climates, I believe. And that is a great future. However, as happy as I was to learn of these new developments, I must remind the reader that we shall always need a portable fuel, in addition to electricity. That is if we plan to continue to have planes, cars as we know them, iron and copper ore purifying, factories powered to do heavy manufacturing, and other applications, like freighting ships, and railroads.

.   .   .

At the level of power plant solar technology, the parabolic trough technology is leading the field. The U.S. Department of Energy [DOE] National Renewable Energy Laboratory (NREL) have collaborated with Solargenix Energy on the solar collector technology to be used in the development of Nevada Solar One, a 64 megawatt (MW) Solar Thermal Electric Generating Plant in Boulder City, Nevada.

"Given today's natural gas prices, combined with tax incentives offered in the recently passed energy bill, utilities and investors are showing a lot of interest in the development of large-scale concentrating power plants," said Mark Mehos, program manager for NREL's Concentrating Solar Power Program. "This is the first in a string of new large-scale domestic and international CSP projects."

"The parabolic trough technique used in this plant represents one of the major renewable energy success stories of the past two decades and has near term potential to compete directly with conventional fossil fuels," according to published information from DOE. Nevada's land and solar potential could produce more than 600,000 megawatts (MW) of power this way. [NREL: Newsroom—October 19, 2005.]

CHAPTER NINE:

THE HYDROGEN FRONTIER

When the value of a fuel is analyzed, just as was done in the chapter on ethanol, one generally calculates the energy put into obtaining the fuel, as compared with the energy that the fuel can produce. This is applicable to fossil fuel input especially. Some people, using this ratio of Energy output/Energy input, say that hydrogen is not a possible choice because more electrical energy goes into the hydrogen than comes out. However, we come upon a logical fallacy here. Though it takes 1.4 kWh of electricity to form 1 kWh of hydrogen*, [Pimentel gives the authorities cited for this ratio in his paper on renewable energies, Bioscience, 12, 2002—available at *dp18@cornell.edu.*]--if we are using efficient wind turbines, currently at 4 cents per kWh, we would have this price lower than fossil fuel simply because once the wind turbines are up and running, they are powered by the wind. At 4 cents kWh, the cost of hydrogen is much higher than it would be, if the wind turbines were mass produced. Probably the most efficient way would be with the windships in the ocean, at 10-18 megawatts per windship, having 36 turbines, each one roughly as high as a six story building. The cost of electricity would go down, to 1-2cents kWh [according to Harry Braun], or maybe even lower. I say this because I believe there would be no "off-peak" cost of wasted hours. The electricity would be as valuable night and day. This may lower the cost in production in itself.

*[It is also true that burning hydrogen in a closed chamber in a local electricity plant is twice as efficient as burning a fossil fuel for electricity. Hence, one happens to be able to get more electricity locally, with hydrogen carrying energy from the wind-produced electricity and that energy used in a local power plant, than one could with fossil fuels.[See Braun, pp. 104-105.] Therefore the ratio, which was 1.4kWh to 1 kWh, now goes to approximately 1.4 kWh to 2 kWh. Though the hydrogen has to be transported to a local plant, in this model, the loss of 50% of the electricity, as heat, along the wire grid is avoided by having a non-polluting energy plant in the neighborhood. I did not mention the kWh involved in cryogenically freezing the hydrogen to a liquid. Dr. Pimentel in conversation said that would cost 33% of the hydrogen—The freezers are electrically run, Mr. Braun said to me recently. At the same site, very cheap electricity would provide the Stirling Motors, "run backwards" to do the cryogenic freezing. Still by wind energy at the electrolysis site, the freezing would be done at the cost of 1-2 cents/ kWh of electricity.

There would be ten men employed in each of the one million windships, Thus there would be ten million jobs created, just in the windships alone. How many subsidiary jobs would there be? Plenty.

The cost, with these windship constructions, catching ocean wind both more steadily and strongly than that of land turbines, and if mass produced, would go down by the efficiency of the windship megawatt range/unit module. As opposed to having 10 million land wind turbines to get the same 100% replacement of all fossil and nuclear fuels, one million windships would be needed, and that off shore 5-15 miles. [Based on Harry Braun's analysis. Some of these figures are found in his papers from the [phoenixprojectfo undation.us] website, and are not in his book.] The lower winds are 5-9 times stronger than on land, and at 500 ft. tall, the upper turbines catch yet stronger winds. The ratio of electricity production goes up exponentially with stronger winds. Winds 4x as strong would produce electricity at 16x the amount, an exponential increase. With wind like that, the turbines would be superpowered, and financial cost to install would take less investment to be prepared to replace all fossil and nuclear fuel use, than with land turbines or the smaller turbines off shore. They would be relatively permanent, once installed. Furthermore, there would be about one trillion dollars worth of consumer energy sales annually, at current rate of use. The energy to turn the turbines is coming from the wind. That part is free. Hence to lose 0.4 kWh in storing wind energy as electricity in the harnessed carrier energy of hydrogen fuel, and 0.3 kWh cryogenically freezing it would not be as expensive as any other fuel, probably. Only coal might be temporarily competitive in price. Sub bituminous coal from Wyoming and other westerly places, does not so cheaply replace the depleting West Virginia bituminous coal. Coal is the greatest pollutant in wide use, with particles, mercury, carbon dioxide and other emissions. Four hundred thousand Chinese develop chronic bronchitis from coal pollution, annually. And it causes the most global warming.

Gasoline to the mass-produced hydrogen cost ratio would be 1.00 to 1.40 by Braun's calculations. [His calculations also show that methane derived hydrogen is more expensive than wind-derived hydrogen would be, in mass-production. This would be the cheapest way to get all the fuel we need. It is necessary to understand that the gasoline at 1.00 is gasoline with subsidies that make its actual cost seem less. External costs make it much more expensive, by far. [Public health costs; damage from acid rain to crops, forests, bridges, buildings, monuments; and military costs, even without an active war, since we have provided a military presence to protect our oil security in the Persian Gulf since the Carter Administration.]

I need to explain to some readers, maybe many readers, that wind is a varying force, so using wind for electricity requires a grid from fossil fuel or hydropower electricity, strong enough and reliably steady, to usefully absorb the varying supply, the unsteady supply from wind. It is generally agreed that a grid or "sponge" of 80% would be needed for wind at 20%.

In addition, one must remember that electricity, as important as it is, is not a total energy supply. Fuel for burning is an absolute necessity for spaceships, jet planes, power plants, steel making, other forms of manufacturing, for freighting ships. Generally, cars need fuel, unless run with the new technology batteries for short distances, as is planned for the Prius +. It is needed for tractors and other farm equipment. In liquid form, or in tanks of metal hydrides that are too heavy for cars, but provide ballast for ships, that permit hydrogen to be released from its bond to metals, as fuel, it is also useful. Compressed hydrogen gas would be used for fuel cell cars, for the people able to afford them.

Hence, everyone who says that hydrogen is not free is correct, but much of the energy to release it from water would be coming from the wind, from a free source to be able to afford to make electricity kWh and transform them into hydrogen the fuel, released from water, by using the electricity induced by turbines turning in the wind.

Subsidies on gasoline make it possible to cost less than hydrogen—But the price of gasoline is very high and tends to get higher, until it is too expensive to extract. But if we put up the windships with lower priced fuel, from supplies that we have conserved, we could keep the hydrogen price down and maybe lower still with engineering improvements. Its cost would have no need to go higher, because there is no shortage of sea water from which to free it, by electrolysis. [There is no place in the atmosphere where hydrogen exists alone, and not in compounds with other elements.] Freeing it from water with wind electrolysis is the cheapest way to get it at present. If there were "fair accounting", taking into account subsidies and external expenses of fossil fuels, as hydrogen starts to be used, that is "clean hydrogen" from water, subsidies could get it started, and there should be a tax on gasoline introduced to give incentive to consumers to buy hydrogen, at a more comparable actual cost, thereby giving incentive to investors to get it rolling in mass production. [See Chapter Ten, "Financing the Hydrogen Frontier."] The question you may have is, can we demonstrate that the windships are seaworthy in some of our less stormy seacoast. The New England sea coast was in the mind of the inventor, stormy, but not like the hurricane zones. The Pacific has storms, but again not like in the hurricane zones. Not many square miles of sea coast would have to be selected in each ocean, perhaps some areas less stormy than others. The windships were built to withstand hurricanes and earthquakes. A patent description indicates that the heavy globe at the bottom, and the largest number of turbines being lower on the structure, rather than higher, gives it stability, even under rough conditions.

The ratio between the 1.4 kWh of electricity input to electrolysis and the 1.0 kWh of hydrogen output, I repeat, does not include the cost of taking a cryogenic freezer to freeze the hydrogen to a liquid for ease of energy density and volume in storing and transporting it. The Stirling Motors, used in reverse to cryogenically freeze hydrogen, are generally run with electricity. There is no reason why initial cryogenic freezing cannot be done at the electrolysis area, also run by the wind turbines. In this way, the cheapest form of electricity would be operating them. No more expensive fossil fuel need be involved. Remember, the hydrogen is always renewable, coming from water and burning to water. If sea based, in a clean, only hydrogen burning piece of equipment, the water could be shifted from the sea to land, and now distilled, pure water could be used for drinking and other uses. There is no impurity in the hydrogen burner to contaminate the water that could be condensed from the water vapor given off in the burning process. The dual tank car, that has used gasoline, whether new or converted inexpensively from one of the 255 million vehicles on the road would not give out pure water vapor, or distilled water. That water would be contaminated with the impurities it picks up from fossil fuel contaminants in the engine. One could not then condense and drink that water, but of course, the water vapor would be much cleaner than exhaust from gasoline. BMW claimed that the air leaving the exhaust would be cleaner than the air entering it in an urban area, according to Braun.

A Brief History of Hydrogen as a Fuel

Hydrogen is the smallest atom to form in the Universe, shortly after the Big Bang which we now theorize to have occurred. As it was very gradually drawn into clouds, in which gravity brought the hydrogen atoms closer and closer together, the point was reached at which it was dense enough to ignite in stars, now creating the furnaces of nuclear fusion that were of such intense heat, that the elements developed, each more complex than the element just lighter than itself. Nuclei were fusing.

It was Abraham Lavi and Clarence Wener, distinguished engineering professors from Carnegie-Mellon University, in Pittsburgh, PA, that were cited in Harry Braun's *Phoenix Project: Shifting from Oil to Hydrogen* [pp. 104-105], that stated that if hydrogen were used for power generation in place of fossil fuels, the costs would be halved. One reason is that the hydrogen, as I've previously stated could burn with air in a closed chamber, making superheated steam that could be fed directly into the turbine. Harry Braun said that this is also true for virtually all industrial processes, directly reducing iron and copper ore, instead of the ore being in a blast furnace, with coal. And since it needs no venting, the 30-40% of the combustion energy of fossil fuels, that is vented as heat and combustion byproducts, would not need to be vented in burning hydrogen. [Braun, p.105]

Hydrogen is extractable from many sources as it is present in almost everything. I do not see a value, ultimately, however in getting it from anything but water, for all the other means have a CO_2 residue. Methane in landfills is suitable for hydrogen extraction, since this methane escapes to the atmosphere, more potent for the greenhouse effect of global warming than is CO_2. It can also be retrieved in animal and human waste treatment, and should be, to fully neutralize the waste material sludge.

The energy systems for hydrogen are modularized, whether windships, or state-of-the art land or off-shore wind turbines, or OTEC boats, or solar equipment when it catches up in durability and cost. This lends them to mass production, adaptation for newer and better technology, incremental purchase and very short construction lead times. Repair is also module by module. Contrast this with the expense of building a nuclear plant, each taking financing for the 10 years or more that it takes to build them. If the system fails, a massive loss occurs, in contrast to the less single-sourced modular systems. In ten years, we could have completed an entire energy system with successful modules of wind production. Hopefully, in less time. The adaptation of fuel burning equipment for hydrogen use would have to be coordinated too, and other countries have been considering the same, especially Japan and Germany. One could hope that the incremental building of windships, started early enough, would fill in the gaps for the energy that we wouldn't have available, some the first year, more the second, still more the third year and so on. There needn't ever be the 10 year time frame of 0 barrels from ANWR before the tenth year, no energy from a new nuclear plant for 10 years, not sufficient Liquid Natural Gas for some time, and an incomplete supply of syngas from coal, for some long time.

One knows of many wind companies or wind-developing nations, including China. They don't talk about hydrogen production. Among the highly trained specialists, there is very little communication. The International Association of Hydrogen Energy is working on many aspects of the problems, but have a range of opinions that varies considerably. There is in a real sense, more acknowledgment of the potential of hydrogen from wind, among the lay people I talk to, than I have seen in scientific journals. I do not understand how people can so easily begin to talk of carbon sequestration, when there need be no carbon dioxide to speak of at all, in a hydrogen future. Government and energy companies, however, can be affected by the public interest in hydrogen fuel, with a vigorous interaction of the American people and the media. Both government and the energy field have been dragging their feet over a long considered transition to wind energy for hydrogen. It was first considered in the U.S. in WWII, when fossil fuel supply was strained.

The Germans and the British had hydrogen powered cars in the thirties, by the thousands, using tanks of metal hydrides which release fuel when heat was applied. This was a heavy storage system for a car and gradually faded out. The same system, though, would give ballast to a ship run with hydrogen fuel, today. Newer metal combinations for hydrides as fuel containers are being studied and published in the *International Journal of Hydrogen Energy* in current issues. This is the official journal of the International Association for Hydrogen Energy, which has members in at least 81 countries.

The Germans also used hydrogen in their submarines and torpedoes during WWII. NASA began with the liquid hydrogen powered conversion of the B-57 in 1956 [Braun, p. 135, Fig. 5.35.] They went on to use the fuel with liquid oxygen in the moon shots and moon landing. The use of the same fuel has been studied by NASA and Boeing-Lockheed in cooperation for commercial and military planes. It is noteworthy that the Germans went directly to use of ocean wind electrolysis of water for their hydrogen, in WWII.

"In order to overcome the problems of siting of wind systems, advanced 'Multi-Array' wind systems have been proposed by Ocean Wind Energy Systems (OWES) located in Amherst, Massachusetts, which has been developing the concept since the 1970's. The concept involves placing multiple wind turbines on a single tower which can increase the power output from an existing site by a factor 10 or 20.

"The principal OWES design engineer is William E. Heronemus, who graduated from the U.S. Naval Academy and worked as a naval architect in the largest ship design group in the U.S. until he retired from the Navy in 1965."[*Phoenix Project,* Braun, p.202] He went on to be Associate Head of Civil Engineering at the University of Massachusetts in Amherst, MA. It was there, that he and associates, invented the Windship that I believe we should begin to put up, in increments. As with spaceships, we understand the principles so all we have to do is put them into effect. Principles of aerodynamics, corrosion, electrolysis, etc.

Roger Billings, according to Braun, developed the first hydrogen fueled automobile in the U.S. It was a Model A Ford truck, developed by the very young Billings, in 1966, while he was in high school. As an undergraduate at Brigham Young University he won an Urban Vehicle Design Contest with a hydrogen-fueled Volkswagen. He, in time, established the Billings Energy Corporation in Provo, Utah, going on to modify a wide range of automobiles, including a Winnebago motor home that was fueled by hydrogen, the motor, the generator, and all of the motor homes' appliances. Then he built an entire homestead with hydrogen. One of his automobiles was a dual-fueled hydrogen/gasoline Cadillac Seville that could change fuels with the flip of a switch. "Billings and his colleagues have been able to demonstrate that

hydrogen could indeed be used as a universal fuel in end-use applications."
[Braun, p.120-121.]

"Because of anticipated fuel shortages, hydrogen research programs have been undertaken by the U.S. Air Force, Navy, and the Army since the 1940's... [but] as long as there was not an emergency, the fuel supply problem was essentially left to the oil companies to worry about." [Braun, p. 123.]

Peter Jennings, quite a few years ago now, interviewed a Midwest farmer, as Person of the Week. With a very small windmill, relatively speaking, this farmer had supplied himself with all the electricity he needed from that windmill, and eventually, all the fuel he needed as hydrogen. He had a sixth grade education. This interview is captured in Braun's videotape, by the same name as his book, *after the credits.* The tape is a very fine production, though it's introductory, technically. My eleven year old tutoring student got a great deal from it. Together we observed the solid rocket booster flames of orange and red, and then most likely, as the flame was large, we could see the usually barely visible blue flame of the hydrogen coming from the spaceship itself. Billings and Tappan stove worked out a way to compensate for the invisible flame, by using visibly very red hot steel wool when it was in operation. My jeweler friend says that his colleagues using hydrogen "water torches" don't worry about seeing the flame. I guessed that they can see the metal heating to the results that they want, and don't need to see the flame itself. He surmised the same, though he doesn't personally use it.

Harry Braun has early information about the development of dual tank BMW's and other internal combustion motor companies. I shall make the next chapter about BMW' s course of research and development, with information that I was given recently, by their public relations and engineering community. Since many of us love cars, when I received the jpegs of their dual tank hydrogen/gasoline sedan, the 745h, I felt a thrill, as at least this car I could afford to admire a lot from a style *and energy perspective.* I saw a BMW yesterday, that had a fuel cap on both sides. They were both for the gasoline tank, but I guess that's a small display of what they will have on the market, soon. I suggest that financing of cars be special, in the purchase of a fine vehicle, not built for early obsolescence. Then the cost won't be beyond reach. BMW plans to partnership with companies all over the world, as a single company cannot fill everyone's orders. The quality in the engine is a necessity. They will certainly not be nearly as expensive as a fuel cell car. We didn't discuss price, though I found out that for about 38 thousand, one could buy a sedan with a lighter engine, "fully loaded", in a gasoline powered car, those now on the road. A pre-owned BMW could compete with Ford or GM, new. I saw one advertised with a 6 year or 100,000 mile guarantee.

In any event, second hand vehicles, all 255 million on the road, can be converted inexpensively—a solution Braun urges, and so do I, though public transportation should be expanded, too, so we clear the road of the long, ubiquitous traffic congestion that is ruining fuel efficiency of any kind of fuel, developing stress, sometimes road rage, and long commuting hours that rob us of the time we could spend with family, once we get the monkey of traffic jams off our backs. That's independence? It is a stress just to *hear* about the traffic tie-ups on WINS news, just before the weather report. I believe that once we got logistically possible commuter transportation more efficient and feasible, we'd never want to go back to the long "parking lots." Extra buses or extra train cars would absorb 25 or more of the cars on the road, per bus or train car. The representative from Greyhound said, "You know, it wouldn't be like getting into your car"—but went on to say that Germany had buses down to a sheer pleasure. There was never a traffic jam. Their buses are also well-designed for comfort.

CHAPTER TEN:

BMW RESEARCH IN USE OF HYDROGEN IN INTERNAL COMBUSTION MOTOR

BMW RESEARCH IN USE OF HYDROGEN IN THE
INTERNAL COMBUSTION MOTOR

I first heard about BMW's dual tank hydrogen/gasoline cars during the aftermath of the attack on the U.S. by the terrorists on 9/11. Braun called in to the radio station I was listening to, and described how little damage would have been done to the Twin Towers had their been hydrogen instead of jet kerosene fuel. He also mentioned the cars that use both fuels. At the time Ford, Daimler-Chrysler, GM, and Mazda were also working on internal combustion engines with hydrogen. Harry didn't preach. He just mentioned these two facts about the availability of BMW's and engineering plans for Boeing-Lockheed. I called for contact information from the radio station, and in time located his book in the 34th Street Research Library. I ordered it right away and had an adventure, reading it. Perhaps six months later I called him—in time we were talking on a regular basis. I remained very interested in BMW's work.

In Munich, BMW has had a hydrogen fueling station available for some time, originally a robotic pump. BMW has been committed to advancing the technology in dual tank cars since 1979. They installed a small fuel cell in the car to act as an electric generator. Their cars looked the same as other models on the road—the only difference was the dual fuel caps. There are no obstacles to mass manufacture as the metals used are common materials, steel and aluminum. There was a Clean Energy Exhibition in Munich, Germany in the year 2000. Their president and CEO spoke as follows:

"We have invited you here so that together we can begin a new era, an era of completely emission-free vehicle power. In short you are witnesses to the fact that we have succeeded in bringing a long-time research goal to a reality. Therefore, I am proud to say the BMW Group has done significant pioneering work in the development of hydrogen power.

Today, we are able to say to you, "Get in—take a seat in our hydrogen car. You can 'experience' the future with us—in the truest sense of the word. What we are showing you in this exhibit is not a study,…What we are showing you is already in small-scale production. If you look underneath the hood of our hydrogen vehicles, you will find that, outwardly, they are not significantly different from those equipped with gasoline— fueled engines. This is what is so fascinating about them. They are completely normal."

This was the President and CEO of the BMW Group, speaking, as quoted by Braun, [pp.327-328], Professor Joachim Milberg.

Professor Milberg went on to say that mobility is a basic need that people have and raises the problem that fossil fuels are limited, and that fossil fuels are related to the compounding of the greenhouse effect. Thus we need alternatives that save energy and protect the climate, as soon as possible. BMW's answer is cars powered by the sun and water. Hydrogen in water is almost inexhaustible.

Mainly due to the high cost and high weight of fuel cells, the group prefers to use a fuel cell only where electricity is used for the auxiliary functions in the vehicle such as air conditioning and heating, when the car is not running.—Therefore, their strategy is to rely on hydrogen power, which provides sufficient power and range—completely emission free.

Braun, who was present at the BMW Exhibition, states that the enthusiasm for a hydrogen powered car and for the more general transition to the hydrogen energy system seems like a part of their corporate culture. There was a fleet of 750hL series driving between Munich—where they built the first hydrogen fueling station in the world that was commercial—and Hannover. There were buses as well. The main difference between their motor and another motor was the additional injector valves for hydrogen as well as gasoline. At that time, they were taking orders for cars—and their fueling station was robotic, taking about 4 minutes to fill the hydrogen tank.

. . .

In the past five years, BMW has continued to develop their cars for hydrogen fuel. The model that will be on sale is the 745h, and the pictures can be seen in the notes on this chapter, both of the 745h sedan, dual tanked, but also the fueling station, no longer robotic. There is no problem refueling, and if there is warmed hydrogen in gaseous form with the liquid, it condenses back to a liquid when the car is filled up, on the liquid "raining" into the tank.

Their website for Clean Energy with pictures of the cars, can be reached by going to [bmwusa.com] to the home page, then clicking on BMW USA which brings up the Site Map. Click on Site Map, and one can find Clean Energy amongst the choices. It is an exciting site, explaining the vision for using hydrogen, plans to partnership with automotive companies around the world, so that the needed production could take place. And they show the race car, H2R, also in my notes on the chapter. It is using only hydrogen, was conceived and developed in ten months, shows the ability liquid hydrogen fuel has to attain power and speed, up to 186.11 mph.

"Concerning fuel efficiency," Mr. Kammerer from BMW, Germany, in a recent email to me through David Buchko, Products Communications Specialist with BMWNA, said, "we can state that the hydrogen internal combustion engine can reach a fuel efficiency of 50% (including an overall thermal management of 8%,) which is comparable to fuel cell applications.

The range of the dual tank BMW combustion engines are from 125-185 miles with hydrogen and up to 300 miles on gasoline, with a top speed in excess of 133mph. In the transitional period, the whole trip across the country would be possible. Where there are no fuel pumps for hydrogen, I am quite certain that a fuel truck, parked, could be a temporary "pump". Eventually, one would need an all hydrogen car, the goal of BMW's extraordinary engineering projections. They are certain that the internal combustion engine will work maximally with hydrogen, more so than it ever did with gasoline. In the long term, the Group's concept is to involve the use of a fuel cell, which will be an Auxiliary Power Unit that will generate air-conditioning and heat while the car is turned off. [BMW Group Media information. 9/2005]

The same media kit states that "Hydrogen has excellent qualities and properties to give the combustion engine a very high standard of efficiency. Apart from the fast rate of combustion allowing optimum combustion management, the hydrogen combustion engine benefitting from its wide range of ignition limits spaced out far apart is able to run on both a rich and lean fuel/air mixture. This, in turn, ensures a very high level of efficiency thanks to unthrottled operation particularly when under part load, exceeding the efficiency of current gasoline and diesel engines."[Ibid, pages 29-31]

The windows and temperatures in which nitrous oxide would occur can be avoided. The area is not necessary for efficient operation of the engine with hydrogen.

Fueling up is no longer done robotically, but is carefully engineered so as to be safe, leak proof, but attached to cables that allow it to be easily handled, weight-wise, by the driver.

At present, liquid hydrogen in the tank will warm up in the course of time, pressure in the tank coming from the hydrogen that has formed a gas. The limit is currently set at 5 bar. *Should the liquid hydrogen grow warmer the tank pressure increasing accordingly, any excess pressure is relieved via a pressure check valve.* And this next statement is important, I think, so as not to have many leaking excess hydrogen in the future:

Boil-off management catalytically converts the hydrogen together with the oxygen in the air to form water. In the next series, the company plans to have a 5kW fuel cell, absorbing all the vented gaseous hydrogen and using it for electrical purposes, such as air conditioning and heat. [I am guessing that it is fair to conclude that the price of the fuel cell is still too high, even at the range of kW the company desires.]

BMW is aware of the need for the public to accept hydrogen, and German polls do indicate that it is accepted neutrally, with some fear that it is dangerous. They have educational media for young people. There is an educational museum in Munich, with lessons on electrolysis, on generation, distribution, storage, and uses. In general the user of a range of educational materials receives a good impression of the benefits offered by hydrogen and becomes more aware of the prerequisites to be put into place by society to turn this fuel into a reality in the future.

There is a Clean Energy Exhibition at the China Science and Technology Museum in Beijing. "To examine the introduction of hydrogen as a primary source of energy in the future in one of the world's largest economies, the BMW Group plans a wide range of activities in China as part of the BMW Clean Energy Project. One example is the close cooperation of BMW experts with partners from both Germany and China implementing a hydrogen infrastructure in the country. Since April 2000, Chinese universities have furthermore been able to use an information kit in Mandarin with the title 'Expert Knowledge on Hydrogen.' And last but not least a BMW Clean Energy internet portal in Chinese makes sure that this information material is fully available throughout the entire country." [BMW Group Media Information, 9/2005, p.50.]

For further information see [press.bmwgroup.com] and [bmwgroup.de/cleanenergy.]

BMW's safety tests:
1.Trying to simulate an accident, one test was performed with full tanks that had intentionally blocked safety valves. The tanks were destroyed under high pressure. The additional safety valves inside the tank, providing for such extreme conditions ensure that the hydrogen is discharged safely, without causing any kind of major hazard.

2. A second test was to have vehicle tanks filled with liquid hydrogen exposed to a fire, completely surrounding the tank, at 1000 degrees Celsius for up to 70 minutes. The hydrogen evaporated in the process being able to escape in a controlled flow via the safety valves. And since hydrogen quickly moves up into the atmosphere, the worst possible effect is that it will burn in a local fire.

Gasoline, in contrast, spreading out on the ground in the case of tank leakage and immediately causing a "sea" of flames when exposed to open fire obviously presents a far larger risk.[ibid., p. 38]

Storage and Distribution of Hydrogen:

Unlike electrical energy, hydrogen can be stored in large amounts in either liquid or gaseous form. This makes it possible to generate electrical energy serving to both separate and store hydrogen. Hydrides, metal powder with hydrogen attached, can subsequently release hydrogen with heat. They are too heavy for automobiles.

Nanofibers or alanates (chemical hydrogen compounds) may prove viable and open up new perspectives for the storage of hydrogen energy.

Transportation of Hydrogen:

Pipelines for gaseous hydrogen exist in areas of high level chemical industry. [In Germany] Natural gas pipelines would be suitable if they meet certain requirements, for example in remaining absolutely tight without any leaks.

[Liquid hydrogen pipelines would not operate with natural gas pipelines, but perhaps could be put up along those routes. They have to be vacuum-pipes and would be made of aluminum. Though an expense, they'd also serve as a perfect electric grid, with no heat loss, a superconductor.]

For liquid, but not gaseous, hydrogen, it is standard practice and could be handled in technical terms, on a large scale by truck. Tanks used for this purpose, as in the case of nitrogen, oxygen, or argon, are high vacuum-insulated double-shell tanks with all the features required.

Courtesy of BMW

Photographs of BMW Clean Energy cars and fueling stations:

1. BMW 745h series, dual tank hydrogen/gasoline, to be sold commercially in 2008. Location at fueling station.

2. Closeup of customer filling tank with liquid hydrogen, at BMW fueling station, conveniently and safely.

3. All hydrogen, H2R race car, breaking records of 184 mph. Conceived and produced within ten months.

4. BMW 745h series, again in fueling station, with customer filling up with hydrogen.

CHAPTER ELEVEN:

FINANCING THE HYDROGEN FRONTIER

FINANCING THE HYDROGEN FRONTIER

The proposal that I am going to make has precedent of significant sort. This country had considerable land grants to colonists from England. Later, we made the Louisiana Purchase of vast lands enabling us to move west to the Midwestern States and some of the Western States. Thomas Jefferson made the Louisiana Purchase with borrowed money. Eventually the winning of the war with Mexico opened up what is now the original 48 states. The Homestead Act allowed people with no land, no farm, voting rights [depended on land ownership--and of course new statehood] to stake claims in the new and vast territory made available, first by the Louisiana Purchase. Now there is no free land left.

Yet many people struggle with minimum wages, high rents, improper housing for a family of good size. They pay the inflated costs of merchandise, fuel, food and other necessities and they have no health guarantees in many cases. What have we left to give them.? A college education with gluts in the employment market for their area of study, career counseling to try to change fields, to other possibly glutted fields of employment. We have a woeful drop off in the birthrate that threatens social security and Medicare in the somewhat distant future. I'd been thinking that that lowering of the number of future workers and consumers will cause our economy to constrict greatly--In the *N.Y. Times* today, someone else noted that upcoming problem.

Ways to increase the birthrate, including among the Caucasians, is important for balance in the future. We certainly don't want slow extinction of the Whites in the Rainbow. However, Europe and the United States are working in that direction. Now, I think of one of the reasons I dreaded the demise of Welfare. Those single mothers with whom I worked were doing a very good job of caring for their children--vastly improved by their eagerness to seek guidance about the well-being of their children in seeking my guidance as a psychiatrist and educator. Their children were thinking big for their futures --we want to be doctors! Some probably made it. Others were emergency ambulance technicians. This was not a welfare-breeds-welfare situation. Mothers went back to school and some took jobs as Home Health Attendants. Making more time available for mothers to care for their children, in any social strata, would encourage multiple children.

Can we claim the wind and the sun for the people here and later around the world, so the earth is the resource of all, not exploited by a smaller number to their enrichment? There are proceeds in large amounts that come from hydrogen fuel if we fully commit to it. From these proceeds we can much more easily pay down the national debt and help pay for the social security system and health coverage. A source of income for the federal government is very important, since the federal government often has to stand by state programs financially. Big income is necessary for a large country. Well, maybe not so big, if we stop our aggressive thinking, our nuclear exploitation, while we ask other governments not to have nuclear weapons.

The Physicians for Social Responsibility are opposed to nuclear weapons and nuclear energy for reasons ethics and health. They are aware of what nuclear damage can do and in 1980 had a film prepared that showed the utter futility of thinking that anything in the path of the wind fire of a nuclear detonation, could stand if a structure, or survive if a living thing. Reagan was asking us as physicians to be prepared for a nuclear emergency, and be available to help victims. That was preposterous. There would be nothing on the landscape. Neither doctors nor hospitals would exist. It was a simple case, on his part, of lack of knowledge to have asked, and very scary to know he and the Russians were both planning "first-strike." Another miracle that people seem to neglect is to give profound thanks for the peace accord between Gorbachev and President Reagan. The next thing on the news was a lot of hollering about "terrorism." Where is our sanity? If peace prevented Nuclear Winter, why did we not all stop and give profound thanks. We don't even have a holiday to celebrate the salvation of that *d'etante*. What is *wrong* with us? That's three miracles that I have articulated, that I heard no one else articulate.

"It is right always and everywhere to give thanks." [Episcopal and Anglican liturgy.] I am just as faulty at this. It surprises me to read that in the liturgy each week. If I were doing it more often, I guess it wouldn't take me by surprise each week and, I, hurriedly try to remember all the things for which I should be thankful. Then forget, quickly. I am so thankful for all the wonderful people who have been helping me with information over the past couple of months and earlier. For the editor who fed me grist for the mill and pointed out that the urgency was more than I realized, pointing me to *The Long Emergency* and the Matthew Simmons article of July 3, 2005. That got me moving, the last, the two in combination, when he pointed out that publishing with a traditional press would put a book like this out of date, production time being so long. He knew I could publish on my own. I was fighting with myself about doing two books on the same subject—it wasn't necessary.

At this point we can only get 8% of their profit for government revenue from the very wealthy corporations. That is a sell-out of corporate responsibility, perhaps the result of the fact that these corporations do function in other countries when they care to, and we are very anxious about them leaving us more stranded for adequate employment here at home. To have the country in control of enough non-polluting energy to replace all fossil and nuclear fuel, hydrogen energy, that is forever renewable, with manageable prices when massed produced, prices likely to go down with engineering refinements, would also have us provided with well over 10,000,000 jobs. The problem of outsourcing would be eluded by the administration being done from within the country. Our GNP will retrieve the 60% that we spend on energy from abroad. Furthermore, we can become exporters of hydrogen.

It would be a show of good faith, if the corporations invested in the bonds we could float for national fuel. The fuel corporations would not appear particularly interested, but they've benefited from their provisions to us, and they certainly could lend to us. Now, my saying this is going to perhaps precipitate a rapid interest in corporation investment, which has not been forthcoming on any large scale. There is more interest in obtaining hydrogen from fossil fuels than solar/wind electrolysis of water. More interest in coal. There is a list of other available fossil fuels in the *Fortune* article I read today, and I know that Dr. Kenneth Deffeyes has been very concerned about the fuel crisis, is one of the scholars who based on geological studies of oil, predicts a peak this fall, by his calculations. We shared an urgency about this when I first wrote to John Kerry, urging him to take interest in hydrogen, when he had won the primary, and Deffeyes told me of a Canadian group who were concerned. I believe that James Kunstler might be connected with that group, who made a DVD, called *End of Suburbia.* Yet Dr. Deffeyes was writing a book about using methane for cars and using shale oil, which is in great abundance when it becomes accessible, according to Jon Birger, in "CRUDE REALITIES: TRUTH AND CONSEQUENCES", *FORTUNE,* September 19, 2005. But one, there's another time delay on getting it, and there's no reason it would be cheap. And it pollutes.

The article states that controlling the price of oil would not only create a shortage as in the 1970's, but might create a line at the pumps and not lower the price. The other consequence of lowering the price would be that the incentive to get more oil, even if from the tar sands, and from shale oil, would be decreased by lower prices. Ugh! More high prices and more technology and more pollution. Global warming is ahead of us! We are behind, the carbon concentration making our Earth a suction for the Sun's heat already. Why try so hard to get alternative oils, etc.? Dr. Deffeyes isn't in the oil or energy business. Yet he wrote a book about alternative fossil fuels, urgently. He told me, "I'll never believe in hydrogen!" It confused me. It was such a contradiction, a conviction of his. A belief system. Simmons and Kuntsler seem to share it. Kuntsler stated so in *The Long Emergency,* apparently not believing in hydrogen as a transported gas, rather than a liquid that was frozen, probably simply because he didn't know about it.

The Schwarzenegger Highway is planning to transport liquid hydrogen for its pumps, which can also release it as a compressed gas for fuel cell cars. Compressed gas was mentioned as a form of hydrogen in BMW's Clean Energy website. One goes first to [bmwusa.com], then clicks on BMW USA, finding the site map. Clicking on the site map you will find many choices. Click on Clean Energy. I got the printouts from their website in May, 2005. As printouts they are not well scanned. However, the photographs that the company sent me are stunning; but go to the website and see and read it.]

If there is any interest from the energy corporations, I don't know if we should accept it, because we get minimal revenue from the corporations [8%], and we would be caught up in a magnatrust probably. Many corporations could invest in bonds for government ownership, if that were possible, but it would probably require at least a three year wait. But things are happening fast. If it were a government project, I believe that it should be conducted with the separateness of NASA at the worst, and the independence of the Post Office at best. Benefits could go for government revenue, but by a constitutional agreement, not to be tampered with as was the Social Security Trust Fund.

The demand for and price of oil will go down precipitously when our demand for one quarter of the world's oil disappears. This could motivate foreign investment from Europe as well as the Orient. And this debt creates sales of one trillion dollars a year in sales. The debt could be paid, guaranteed. Hopefully, we can assist other countries to enter the hydrogen frontier. Pollution does not start and end in America. The world need for improvement remains. Hopefully, China can afford it. Any excess from America or maybe even from China to countries who have no access to ocean places and have poor sunlight, could be sold at the same price as we pay for it domestically, though in time probably most countries can supply themselves with wind in their deserted areas. I don't believe that we could ever run out of hydrogen there are so many natural solar/wind means to get it eventually. Why bother with oil?

One of Ross Gelbspan's best suggestions is the thought that Saudi Arabia [and maybe other Arab countries] could be bailed out of their sudden economic collapse, if Israel, who have "cutting edge technologies for solar energy", helped out Saudi Arabia, [whose solar techniques are not yet as good as wind, from my reading in a Saudi newpaper.] [*Boiling Point,* Basic Books, 2004, p. 190.]

As it is, it would take a New Creation, so vividly described in the Book of Isaiah and the Apocryphal book, the Wisdom of Solomon, to take the natural deadliness out of nuclear wastes and out of our negative carbon-cleaning balance and our negative heat balance, as far as I can ascertain.

We have to try, however. We cannot just further provoke our atmospheric heat and ocean heat problems, knowing what we know. I remember a couple of natural miracles. The first was in the winter of 1979-80 when fuel was expensive. By what I observed it turned out to be a mild winter.

The second miracle I have a more detailed memory about. There was a gypsy moth plague on the trees of the Eastern seaboard. No one wanted to spray, remembering Rachel Carson's *Silent Spring*, lest the birds be killed. They didn't spray. Sticky tape was wrapped around the trees, not doing any good. The caterpillars could be heard crunching the leaves. The mountain top was empty of leaves, in August, and it looked like November. My father read a lot and was very grieved. He felt close to nature, if not to God.

He had explained that these moths were imported from France with no natural enemies in North America. They hoped to make silk from their cocoons, the motive for importing them. He told me this years before the plague occurred in 1980. In that terrible plague, he said that the trees would die in three years, [enough time to go across the woods in Canada, and onward to any deciduous trees.] I silently thought, well if we sprayed, fewer birds would die than if all the trees die that were leaf bearing. I didn't say anything. I guess I didn't because people refused to spray with what was like a religious conviction. I held my tongue. The next year *there were no gypsy moths* in western Massachusetts.

I said to my father, What happened? He answered, a virus developed. Science and religion came together between us, unspoken. We both knew this was a miracle. For a virus to develop is common, but for a virus to exterminate almost every last gypsy moth was a miracle. How to get a natural enemy in a hurry!

So I kind of feel that if we solemnly, sadly try our best with our biggest enemies, using the intelligence, devotion and common sense that God endowed us with, before we each fell, the shortage of fuel and an already overheated greenhouse we call Earth, maybe we'll find ourselves in the midst of a miracle after we put our foot into the Red Sea water. It is a tradition to say that the waters of the Red Sea did not part until the Hebrews had put their foot into the water, pressured from behind, and maybe trusting in the truth of Moses' messages from God, a little bit.

Now suggesting that the government own the energy industry, the "hydrogen frontier", will bring out cries of socialism. We are capitalists. The market economy is our "sacred" rule. But is it at all sacred? It is full of ingenuity and products that could never be planned by a bureaucracy, but it is full of unplanned ups and downs, and opportunities for evasion of financial responsibility to stockholders, let alone the country which gives it the opportunity to be. It has no mind, no plans for justice, no plans for stability when facing emergencies--and it has been spread throughout the world with no real responsibility to its employees in many cases. It favors ownership alone--that is the only guarantee. And we're feeling the effects of it throughout the world. It certainly is not God, yet we talk about the market economy as though it is "sacred", not to be questioned and not to be contradicted. There is an excellent article in *Christianity Today,* September, 2005, on the importance of business as work and ethical responsibility to customers.

Well, we are now in a situation where the market economy has failed us. Because the price of drilling oil was high and the chance to retrieve much was low, we have no new fields and reservoirs drilled. Drilling only became worth it recently. And you cannot develop a well overnight. Granted we wanted to drill ANWR; forces were against it. So just now, we are ten years away from getting production from ANWR if we do drill it. Last spring we wobbled around the supply-demand equation. What do we do in the gap? [Much of this chapter was written in the spring of 2004—It's a good vestige from my original manuscript.]

We have a chance to conserve enough to institute a pollution-free alternative, a 100% replacement of fossil and nuclear fuel. The mechanics of doing so are clear in the mind of a very ingenious analyst of the decades of major research on wind/solar hydrogen fuel. No one is as well thought out to inform us with his writing and analysis as Harry Braun--granted his plans may not be considering every vicissitude--but he has almost every point considered, engineering-wise. The bulk of the discussion I make about hydrogen, comes from what I read and discussed with him. I have had to learn to go on without him, and learned more, saw more angles on my own. But he could teach and answer most questions and participate in discussion with others who are interested in developing the hydrogen frontier, such as the patent owner, the engineers who work with her, with me, and with others.

Since hydrogen fuel is a fuel that can be used in any kind of fuel burning equipment one can think of, and has been used a great deal in the past, it is not to be neglected. It's an amazing alternative--it takes no land, if done as I described it with windships; it uses rising seawater; it desalinates water in some designs, it burns to water that can be either collected for drinking, irrigation to clear out mineral salts destroying agricultural land, can add to depleting land water; and it can help relieve the atmosphere of global warming gases and pollutants from all our current fuels. It is forever cycling from water to water, so can never run out. It has properties that can greatly reduce our electricity gridding problems. Located in the sea, it can protect fish where there are windships and can feed a rich stock of shellfish where there are OTEC ships pulling plankton up with the cool water, which upon mixing with warm tropical pools, enable turbines to turn and generate electricity 24 hours a day. The income from the shellfish alone would pay off the infrastructure. This would be in the world's solar heated pools. [See Harry Braun at website and in his book, *Phoenix Project: Shifting from Oil to Hydrogen* ,2000, SPI publications, at [phoenixproject. net]. Just using the windship plan would create well over 10 million jobs, raising our GNP considerably. Proceeds could go to social security and Medicare.

I read last night that the new Medicare law allows for *rationing* of life-saving medical care. Old people will be like Eskimos gone off to die on ice cakes under this system. Built-in was the proviso that people who could afford to, could pay for the life-saving procedures, but at this point, I couldn't! Could you! Maybe our federal income from the hydrogen frontier could eliminate involuntary euthanasia, dictated by United States law. With hydrogen fuel profits and decent corporate tax participation, our human needs could be taken care of in this country. "Social ownership" of this one industry would not destroy democracy, but enhance it. If people receive questionable entitlements because of inadequate income from corporations, with borrowing for war, the expenses of which go from left to right on the graph, all the way to the right. Expenditures for everything else are just relative blips on the left of the government's graph.

[See letter to Warren Buffet, where I raise the question of financing without the government, through a Citizens' Non-Profit Corporation, perhaps to in time be incorporated by the Federal Government, independent of political whims, like the Post Office is. Perpetuity might require this incorporation by the right climate in the Federal Government.]

In terms of buying profitable Treasury bonds, which are <u>not deficit financed,</u> [Mr. Braun states that if the government financed it, it would be deficit-financed.], there could be a healthy return on investment. For, if owned by the government, there would be public profits from the energy infrastructure investment. Many other aspects of this totally integrated economy could be private. The more interested investors might be willing to invest, without fighting this social move. They may be willing to pay more taxes. Something has to be done to have corporations and banks working with the people and not only for their own futile empire building. They are suffering a disease to dissociate themselves from the people of normal means or little means. The bonds could be attractive economically. We are at a point where clearly so many think that life is about making immediate, and large profits. I received three letters from millionaire clubs recently to entice me into the notion that having the biggest yacht--What's so great about a yacht? I enjoyed sharing a day in someone else's, but I have dozens of things I'd rather do than worry about the upkeep of a yacht. Right now, I'd rather go horseback riding on any decent horse I can make friends with. I did that when I was younger, and the stable horse saved my life, by swerving off the path as my head was about to get to an overlying tree branch that could have killed me. Moreover, I managed to stay in the saddle. The whole enterprise was very inexpensive and exciting. But I'm writing now, glad to have my own computer, so I can get this job done. And deeply satisfied that I received an unexpected loan to pay all the necessary expenditures to have this book on the market properly.

I love to spend my Sunday afternoon at my evangelical Episcopal church. We're having a spiritual adventure there, together. The pastor knows about my concern for the world's protection, both the earth and its inhabitants. It was the thought of hordes of the people I've come to love, here in the City, and elsewhere, that most strongly caused my dread of the fuel crisis. I had already decided about five months earlier to try to popularize the crises of planet Earth. [On Earth Day, 2001] In December of 2001, I learned about the prediction of a fuel crisis, but not till 2010. It started building up in 2004.

We can't force corporations and banks to invest, but if they don't find it within their hearts to do so, there is an advantage to other countries doing so, who can't go hydrogen themselves, right away. Our conservation, then our hydrogen supply, will lower the cost of oil around the world. The next step is to get the program activated in other countries. China could start at the same time, I'd suspect.

Going back to corporate cooperation, if there's any Christianity involved in these decisions, we just need to recall which letter it was of St. Paul's that stated (l) his support money should definitely be there when he arrives and (2) the people who had more money should contribute more. There have always been people who had more money than others. This was indicative of capitalism. Their contributions to the church were supposed to be each according to his benefits giving to those who are in need. Originally, in the book of Acts, as the Church was first forming everyone was selling property for the common good--you will remember the husband and wife who tried to cheat and hold some back for themselves. They both were overcome with death, it was reported.

MARY ANN T. SEGAL, M.D
NEW YORK, NEW YORK

Warren Buffet
Berkshire Hathaway
1440 Kiewit Plaza
Omaha, Nebraska 68131

September 13, 2005

Dear Mr. Buffet:

Given your rare and celebrated talents in the financial field, I am writing to you to consult about a rather unprecedented economic construction, when all constructions of the problem would be unprecedented, I believe.

I am completing a manuscript, tailored to current emergency conditions, called *Getting through the Wilderness: The Fuel Crisis, Global Warming, and the Hydrogen Frontier.* "Wind power is currently the most cost effective renewable power technology. In conjunction with water electrolysis, wind can provide hydrogen, at the consumer site with few emissions and with very low consumption of fossil resources." according to the DOE Hydrogen Posture Plan, p.A3, February, 2004.

As I understand it, liquid hydrogen (and sometimes other forms) is usable in any fuel burning equipment, directly as a liquid in most applications. This is not to mean that fuel cells are being used. The hydrogen is being used directly to burn to water vapor, with pure oxygen in the NASA spaceships, but with the oxygen in the air, within the atmosphere. The Germans were using hydrogen in submarines and motor vehicles. NASA first used it in a jet fighter in the '50's. Fuel cells were used in the cabin, but of course, do not provide the very high energy thrust to get to the moon, and eventually to Jupiter.

Boeing-Lockheed has engineering plans for liquid hydrogen fuel, the take-off weight being 40% less. It is light, powerful, and safe. [Both NASA and BMW (with dual tank gasoline/hydrogen internal combustion motors) have done crash studies and find hydrogen less dangerous in a crash than any other fuels that we use.]

In the World Trade Center disaster, hydrogen equipped planes would have been very much less destructive, as hydrogen quickly dissipates upward into the air, the molecules being so tiny and light. Water vapor is the "smoke" of hydrogen combustion, so there would be far less danger of smoke inhalation for firemen and people in the situation. High octane jet fuel splays out sideways, sticks to the skin, and burns very severely. The lack of dissipation of fuel, thus made the fire in the World Trade Center, very ferocious in heat, and so actually melted the girders of the Twin Towers.

BMW's well tested and carefully developed cars, begun in 1979, are ready for sale, within the near future. They seek to partnerships all over the world for the sharing of mass production, anticipating that the hydrogen economy would be very healthy. Harry Braun discusses cars on the road, being converted to similar dual tank cars, at low expense. This provides for the possibility of total rapid transition to hydrogen use. Fuel cells are not in the picture, except to provide a small 5kW fuel cell, to capture vented hydrogen from the tank, where some has warmed, and thereby making a battery for the car, where heating and air conditioning is possible, even if the car is turned off. I have no reason to doubt that Mr. Braun's statement that the 255, 000,000 vehicles on the road could be adapted for hydrogen with promptness. Then we would have a fleet that can use hydrogen, in a new BMW or affiliate car, for those seeking new cars, and a fleet of vehicles that have been and will continue to be on the road for awhile.--In effect, this transition of cars is truly affordable. The scientifically beautiful concept of a fuel cell is fascinating, but obviously far from being practical for broad use, due to very high expense. Yet most people have not heard of the BMW's use of hydrogen in an internal combustion motor, just as clean as a fuel cell. They have a racing model that's won records in Europe, going 184 miles per hour.

Meanwhile, we are communicating about the implications of their statement for the general public that hydrogen is 3X's as powerful as gasoline. In engineering terms, it is stated more accurately, and I await further elucidation of the bottom line implications for cost, according to these statements. The answer, simplified, is that by density hydrogen is stronger than gasoline, but that gasoline has fewer liters for BTU's than hydrogen. Therefore, hydrogen has more energy in density, but not in liters. A 4 liter supply of hydrogen is equal to a 1 liter supply of gasoline

The information I'm sharing with you is not widely known, as fuel cells have dominated the literature, and all sorts of comments are made about hydrogen, from people with little knowledge of what I have learned. As I understand it, liquid hydrogen can be burned in a combustion chamber, getting superheated steam to efficiently turn turbines, so 50% of the capital cost of a power plant is reduced. [See p. 105 of the *Phoenix Project,* the 2000 ed. of the book by Harry Braun.] This is a 50% capital cost reduction over using fossil fuels.

As natural gas depletes, where we need more energy in the sponge for being able to use 20% wind energy, a goal for a lot of Americans, we must inevitably return to using coal –the only thing I know of to plug in where the coal would start to dominate again, is liquid hydrogen.

Without it, I envision coal, coal and more coal. Not irrational if hydrogen weren't a clean and adequate substitute, but very unfortunate for the pollution we want to eliminate, if hydrogen is a clean and adequate substitute. *Fortune,* February 21, 2005 had a depressing article about the coal that seems inevitable. It was written by Jeremy Main. It would be ten to 20 years before the new techniques for its use were developed, including syngas. Widespread use would necessitate billions of tons of carbon being sequestered. Hardly sounds possible or desirable. Then, after 200—250 years, the coal would be gone. And so would human history be gone, as we know it. By then there would be no oil, probably no natural gas, and not even coal. People might be survivors, but the population would be greatly diminished. Wood and fiber plants would be the heat source and cooking source, I suppose.

I just realized in the past few days how close we are to starting up a long and never ending coal electricity program. I didn't realize it was so close. It prompts me to highlight the benefits of hydrogen, which is forever renewable if it comes from water by solar, wind, and other clean means.

We toyed with hydrogen in our administrations over the decades. It was never cheap enough. We went to nuclear plants, thinking they would be a much less expensive way to get electricity. Yet that turned out not to be the case. Well now hydrogen from renewable sources is essential and at the eleventh hour. I say in my book to conserve oil as massively as we are able, to get the price down for real necessities, to lower emissions in the process, glide more slowly with the supplies of oil that still exist at easy extraction rates. [Refinement rates are now an added problem.] I also say let's get the price down for the infrastructure of hydrogen. We can give it a year, because no one will act in a year anyway, maybe two years—but it is urgent enough to get it into mass production by wind techniques to start. They are considerably cheaper than solar techniques, though a promise of a great breakthrough with NanoSolar, Inc. is being researched. It may get the cost down 99% over photovoltaic cells, and the efficiency up 500% over PV's. That could unleash the solar sites around the world, including in Saudi Arabia. We'd have magnitudes more energy available.

However, at this juncture, we know that wind electrolysis of water is the most cost effective route. And I know of a system that seems exceptional, using wind ships, masts of hexagonal shape with 36 turbines, reaching 500 ft. into the ocean air, and they are 250 ft. wide. The number MW of energy for each wind ship ranges from 10-18. Land turbines are much lower in MW capacity. One million wind ships would produce 100% of the energy used by all fossil and nuclear fuels in this country, at this time. Harry Braun has estimated the cost to be three trillion dollars, or so, for full production. The same amount for land turbines to produce the full energy supply would be 10,000,000 land turbines, costing more by a factor of 3-5. The opulence of the sea would place those Naval architect and engineer's wind ships in a 25.5X25.5 mile area, split probably between the two coasts That's not a lot of sea space. Ten million turbines on land would be a clutter, comparatively. One would also have to bring the water to the land turbines, but not so at sea. W.E. Heronemus invented these windships in the 1970's to make hydrogen off the coast of New England. They are undoubtedly very sea worthy. In my losing communication with Mr. Braun, who shows the wind ship on his first page, at phoenixproject.net, I gained communication with the woman who owns the patent rights to her father's invention. Some improvements have been made. I think they are the way to go first, and can be put up in modules, testing them as we go. A move to get them fully functioning within five years is in order. Use of the hydrogen would put an end to further pollution, an end to further global warming, and an end to the fuel shortage and depletion. I believe China's state money and massive labor force could perhaps use them, too.

They have considerable seacoast and little is needed actually for our 100% fuel shift. They could increase their energy supply without pollution. They do have more wind turbines up and running than any other country in the world. This information came from a Chinese website, [china.org.cn.]

Well, whereas a large consortium of energy companies would have to finance this, in effect a large trust, with possible motives to continue with coal, satisfied to get "dirty" hydrogen, and sequester carbon dioxide—not really forthcoming with the clean wind hydrogen, with one possible exception of a land turbine project in the West, to the tune of 5% of national needs, for 500,000 land turbines, the cost proportionately higher, I think of alternatives to the energy companies being sole investors.

There are two other alternatives I can envision. Since hydrogen would have vast sales, at 100 Quadrillion BTU's, a bit more than we use in a year now, it is a money making proposition. The government could sell bonds to myriad investors at a decent % return for their investment, without being a deficit financed project. But the current government is not interested in any proposal that doesn't potentially leave us in a significant energy gap. ANWR for however many barrels it contains, would take ten years, and a half of a trillion, to develop, before one barrel of oil would be available. Nuclear plants, not desirable, take ten years to build, require lots of fossil fuel to build, create more nuclear waste, and may not see profits for 17 years after they are built. And how many would need to be built to get 100 Quadrillion BTU's of energy per year? Fuel cells are the administration's goals. Many agree that fuel cells are too expensive for general consumption. They are a vested interest now with all the money put into their development. They just won't run jet planes and will be pretty overworked running freighting ships and 18 wheelers. We are already using a lot of hydrogen, getting it from methane feedstock at higher prices than from wind turbines. [Mr. Braun's very conservative cost analysis states that we could get hydrogen for $1.40 compared to gas at a $1.00 baseline. He does point out, however, that this is subsidized gasoline, with multiple external expenses that we pay as a country of taxpayers. Public health costs, crop loss from acid rain, damage to buildings and bridges in need of repair, military protection, etc. Gasoline costs much more like $5.00 at its baseline of $1.00, perhaps we could say much more. So that hydrogen figure is cheaper than gasoline and could go down with improved technology but never up, as it is forever renewable.

But the present administration is not interested in a clean hydrogen investment, and a new administration might or might not be either, as I have not heard much about radical change, but primarily improved CAFE standards and wind among the Democrats, other than myself. I _am_ a Democrat.

Yet investment in this infrastructure, I say that I favor at this time, and consider urgent, is an investment that would greatly increase our GNP and provide a myriad of jobs. There would be new appliances for hydrogen use, akin to the inventions people loved in the early 1900's. Oil would phase out, leaving some for the many things we manufacture from oil. If the NanoSolar, Inc. microscopic, three dimensional plastic polycrystals work as hoped, the sun countries could begin to reasonably make hydrogen, probably cost competitive with wind. OPEC countries could participate. Saudi Arabia has been thinking solar for a long time. We could aid them where necessary as income drops off, for a country that has essentially one industry, and lots of dependents who get welfare from that one industry, and who will need some first aid, perhaps.

Back to the mechanism of investment. Here I propose something that is perhaps unprecedented. Could we consider having a *Civilian Energy Board,* composed of qualified business managers, and engineers, or others who learn or understand enough about hydrogen, to be able to contract the work, make appropriate arrangements with the patent holder of the Heronemus Wind Ship, form a constitution that is non-profit for board members, who would be salaried, paying some parties who are knowledgeable enough to be of great educational and technical help, but perhaps more profit-seeking, the salaries they'd accept if they cooperated, and if the board found them useful, but basically keeping expenses in the zone that would not inflate the costs of hydrogen for the people. There is no reason why the cost of hydrogen should ever get higher. Improvements in technology might lower the price. In turn, profits could go to agreed upon outlets, one of which might be to help a country to achieve some energy independence who could not otherwise afford to, to add animal and human waste treatment as another means of hydrogen supply, at the same time cleaning the rivers that land on the continental shelf, poisoning the ancient microbes that supply 60% of our oxygen and create ozone, give the general revenue of the government a percent of the proceeds for targeted needs [health insurance?], certainly more than the energy corporations are contributing to the federal revenue. More taxes would be coming in from the contractors, and the myriad new workers. Ten million, just to man the windships.

Bonds then could be sold to many banks, corporations not in energy, corporations in energy, Wall Street investors, individuals, at a decent rate of return, perhaps payable in 15 years. The housing investment is soft now. We have exploited the oil investment. Wall Street Week about a year ago said just about everything is figured on already. A percentage matching the recent mortgage rates should be sufficient. Bonds would not be sold until business and manufacturing plans were realistically solidified. Private entrepreneurs could pick up the manufacturing of all the subsidiary aspects of this. This Citizens Board Nonprofit could be the management core to launch the hydrogen frontier, raising the money from the bonds to pay for the contractors and materials, etc.

If perpetuity looked like a problem, perhaps the business could be incorporated by an administration who were prepared, perhaps being an aspect of government at least as independent as the Post Office, so that whimsical political interference did not occur.

I would very much like it if you would give me your comments about the possibility of such an economic construct as a non-profit Citizens Energy Board, given that investment is large, and there is no way to avoid a magnatrust of energy people, whose motives could be other than to complete the project to its highest potential, for the benefit of the consumers. There has been lack of responsibility in the energy field, if nothing else than to permit Detroit to turn out gas-guzzlers at discount, up to the day of Katrina's $70 a barrel tragedy. I believe that there must be investors to start the ball rolling, who really could worry with me that more and more coal is our fate in the electricity field, and less and less affordable fuel in the transportation and heating fields. Until we may be lost.

Sincerely yours,

Mary Ann Segal, M.D.

CONCLUDING REMARKS

CONCLUDING REMARKS

In a news article from the Associated Press on November 16, 2005, there are indications of a change of attitude in Congress about efforts to stem America's appetite for oil, so much imported. There is a push from environmentalists, evangelical Christians, and political conservatives acting in unison. The conservatives are responding to the political insecurity of depending on foreign oil. The goal of a bipartisan group of senators is to save a bit over 10 per cent in a decade. The legislation includes tax breaks, as much as 35%, and loan guarantees to get auto makers to switch from gas guzzlers to gas-electric hybrids, advanced diesel, or other alternative technologies. There would be tax breaks for people who bought such cars and vehicles for fleets, for ethanol development from cellulosic plants, and promotion of mass transit corridors. This is where the public wants to go and there's been a major change in thinking since gasoline hit $3 a barrel.

There was nothing about hydrogen as a fuel, specifically, and fuel cells were not mentioned either.

It is good that there is a change in mental sea. There is no indication that what I have shared in this book is being thought of or is known about as of yet. There is no request for urgent fuel conservation, per se, on the grounds that our current demand cannot be decreased immediately without effort to do so, given that many SUV's and 18 wheelers are on the road, and that excess driving has been contributing to the high cost of fuel. There is no sense that in ten years, a 10%+ decrease may well be too little, too late for the tentative foreign fuel supply, and that 10%+ reductions in a decade will have little impact on global warming problems. In effect, it shows a smorgasbord of ideas, like cellulosic biofuel that is probably energy negative and very little fuel in amount, that even if energy positive would not help a plane or factory function. It takes a great deal of fossil fuel to create it. According to Pimentel and Patzek (see chapter five), there are chemical and physical factors that make biofuels unable to be energy positive.

Well, be that as it may, there is a change from the earlier rigidity in facing the same kind of legislation.

What I should like to do now is frame some statements that my readers and those who understood and thought there may be much in what I say, could present to the Congress:

1. We have become aware that the price of oil is at a constant increase along with the demand, worldwide, even now while the price is painful. The

demand is of smaller increase this year than usual, but it is still higher than last year, world wide. We ask for a nationwide effort to conserve the use of oil now, to contribute to an immediate decrease in its use in the U.S. and a consequential lowering of price for real necessities. Many Americans are finding the price of gasoline, electricity, and heating fuel out of reach when all put together. As for natural gas, we understand that the domestic supply is depleting and the expense of importing Liquid Natural Gas is high, and that we need wind, solar, hydrogen supplements quickly to offset our shortages of fuel and electricity, especially before more coal is brought into electricity plants to replace the diminishing oil and natural gas. Our citizens are hard pressed to pay energy bills at this time. There is considerable short-fall of money for heating.

2. While considering alternative technologies for automobiles, we want a greater clarification of the fact that internal combustion motors can be fueled inexpensively with liquid hydrogen, whether new, or already on the road, using dual tanks for hydrogen and gasoline. This could open up new supplies of clean fuel at the fastest rate we can handle.

3. When hydrogen cars are discussed in Washington or in the press, they are almost invariably fuel cell cars, for which great expenditure of research is going on. As we understand it, they are very expensive and will not be in the price range for the common individual for the foreseeable future, if ever.

4. It could and should be well known that BMW has developed competitive hydrogen technology for the regular internal combustion motor that would make the price of using hydrogen sensible and reasonable, available to people wanting a changeover to hydrogen fuel. Their cars are safe and environmentally pure. Other companies have developed them in the past, but no longer speak of them. Following the principles of the new dual tank BMW's, second-hand vehicles can be quite inexpensively converted, and effective.

5. Hydrogen is most inexpensively derived from wind generated electrolysis of water, according to the DOE Hydrogen Posture Plan, 2/2004, A3, and that technology should be tried and tested with already long available technology, developed decades ago, but not used for "market economy" reasons, or because it was erroneously thought that nuclear energy would be much cheaper. Land turbines can generate hydrogen by electrolysis of water, but the multi-array windship of the Naval engineer, W.E. Heronemus, would take up no land and little seaspace, and be far less costly for mass production. We want Congress to encourage the immediate construction of some prototypes of that invention, and use them as much as possible, as soon as possible. We can create a total replacement of global warming

fossil fuels with the use of hydrogen, long studied by our own government through NASA, if this technology proves successful. Built in increments, over five years to ten years, one could make engineering improvements if necessary. One hundred percent use of hydrogen, in place of fossil fuel is possible. It is the only way to be certain of having enough fuel for all our needs and to have it be totally environmentally friendly. Water, not carbon dioxide and other polluters are emitted when hydrogen is burned. It cannot run out—There are magnitudes of sources to make it in the world.

6. To consider obsessional research about hydrogen as a way to postpone its use, guarantees that we shall be in an energy gap, while waiting for new and expensive fossil fuels to replace our diminishing supplies, only to heat up global warming beyond the breaking point. Or in simply substituting global warmer villain number one, more coal, to ease what is now a crisis with natural gas.

.   .   .

You may not believe all this yet. For example, do you have a car on the road that I am claiming can be safely and inexpensively converted to a dual tank car, the second tank under the backseat, with fuel injection lines for hydrogen going directly into the cylinders, and bypassing the carburetor if you have one? Harry Braun says it costs 3-4 thousand dollars and that you should be subsidized for converting your car, with money obtained by a tax on gasoline, at the right time, to encourage the use of and investment in hydrogen. He calls it a Fair Accounting system, to level the playing field of subsidized gasoline. It used to sound harsh, but as the developments of oil shortages have taken place this year, we see the government trying to subsidize gasoline some more. The energy companies profits are higher this year, so there is more going on than a simple increase in price for an over demanded supply. That profit level should have come out even, if it were a simple supply/demand problem, except for the manipulations of the players in the oil field.

I want to refer you to Mr. Braun's H2 Fleet site at [Phoenixprojectfoundation. us], the Hydrogen Fleet site. Mr. Braun wishes to raise enough contributions to have a fleet of dual tank hydrogen/gasoline cars, BMW's and second-hand conversions, along with a large bus that is a mobile home, completely run from motor to all its appliances, with hydrogen fuel. This motor home illustrates that hydrogen has many end-uses, is a universal fuel. This would begin to graphically illustrate that we could quickly convert all the cars on the road, along with new purchases, to be running on liquid hydrogen. This would enable a very rapid transformation from fossil fuel to powerful, effective, and forever renewable clean energy on the highway. And with further illustration of effectiveness of hydrogen in many other applications, a rapid transformation in our electricity plants, and many other uses of it that can be made. This transformation of fuel is essential to our survival both of the cold and breakdown of affordable mobility; there could be a breakdown of the food supply that endangers civilization. Hydrogen's rapid and wide spread use could put an end to the further aggravations of global warming. It is not science fiction or rocket science. It is simple common sense. Harry has a plan to display these dual tank cars, the hydrogen independent large bus/motor home, and the fuel tanker that will keep them supplied with hydrogen in every major city from California to Washington, D.C. These are already established prototypes for some long time—They are not widely known, because very few have been properly exposed to this information. Seeing is believing and there could be adequate media coverage.

I disagree with him about a few subjective things, but I have found his technological insight ever fruitful when I read what he writes. I don't agree with his politics and religion, but that is neither here nor there. He is a genius and immensely informed, aware for a long time that if we did not take the hydrogen route, we would die off, after ruining the planet. He is not unique to looking for the hydrogen economy, but his very practical notions show how we can move with the speed of a Manhattan Project, as we did in WWII.

Harry did not see why I was emphasizing conservation—I think he saw me as deviating from the path we'd established about the hydrogen information, without realizing that I was worried that we'd deplete our fuel and hamstring ourselves, before we could get the infrastructure up for the hydrogen.

He probably would not agree with my precepts about funding the Hydrogen Frontier. He thinks that energy companies should fund the energy changes. But I'll leave him to comment about that, if he cares to.

His hydrogen fleet idea requires funding and one can vote for hydrogen by making generous contributions to his H2 Fleet plan so that it will be vivid and clear that such vehicles exist and can function very well to make our transportation continue safely and free of everything but water as a byproduct of burning the hydrogen.

.   .   .

I hope that some of the cities I reached by press releases in newspapers across the country had a chance to think about heat shelters already. My anxiety was matched by that of the government official with whom I spoke in Maine. Heating oil was reported to be 27% higher than last year by the government on November 8. I believe that natural gas is still proportionately higher and that electricity is higher as well. Con Edison in New York reported to me that natural gas would be about 71% higher but that they had locked in an arrangement for electricity to be only 30% or more, higher than it is now. October was twice the price of the same month last year. It still seems like many people cannot afford their heating bill this year, with the little that LIHEAP provides. And many others who cannot afford heating this winter, are not eligible for LIHEAP.

The last report that I heard was that Congress failed to raise the heat assistance and that eleven Democratic Congressman wrote to nine oil companies, asking for some sharing of the profits they made, to assist low income people. Only President Chavez of Venezuela responded through Citgo. He sold some discounted oil to energy non-profit agencies in the Bronx and in Boston.

If heat shelters are not being quickly envisioned, the possibility of getting together with two or three other households, sharing the heating bill as house-poolers may be more in the control of individuals who are pressed. The Government official in Maine said her elderly population found that more palatable to join friends and relatives. Maybe that's the best we can do until the mechanics of larger shelters can be worked out. Housepooling cuts demand in 1/2 or 1/3—There's reserve built up with that too, on a smaller scale. To this point, no one has expressed responsibility to pay for the heat in the heat shelters. This is a very new emergency pressing upon us. Last year caused grimaces. This year would cause danger. The Red Cross is available if there is a particular cold snap emergency.

. . .

In church this Sunday there was a tiny girl, with tiny black boots for the snow. I looked at her boots. They were enchantingly tiny for leather boots. It had been a beautiful service. I felt very certain that she would survive to maturity.

BIBLIOGRAPHY

BIBLIOGRAPHY

BOOKS

Braun, Harry—*The Phoenix Project: Shifting from Oil to Hydrogen,* 2000,
      SPI publications, Phoenix, AR
Gelbspan, Ross—*Boiling Point,* 2004, Basic Books, New York, NY
Goodstein, David—*Out of Gas,* Norton Books, 2004 New York, NY
Holy Bible, New Revised Standard, with Apocrypha, 1979,
      Oxford University Press, New York
            King James Version
Kunstler, James Howard—*The Long Emergency,* 2005, The Atlantic
      Monthly Press, New York, NY
Roberts, Paul—*The End of Oil,* 2005, Mariner Books, New York, NY
Wallis, Jim—*God's Politics: Why the Right Gets it Wrong and the Left
      Doesn't Get It,* 2005, Harper Collins, San Francisco, CA

VIDEO AND DVD

Braun, Harry—*The Phoenix Project: Shifting from Oil to Hydrogen*
            Video Tape, available at Phoenixproject.net
End of Suburbia—Matthew Simmons, Kenneth Deffeyes, and others speak
on this DVD, saying the rush to Suburbia over the years, has put us too
far from our jobs, with depleting oil. [endofsuburbia.com]

ARTICLES AND WEBSITES

Chapter One: *Observer* , 7/3/05, Matthew Simmons states "Oil May
Hit $100 a Barrel in Six Months"
New York Times Magazine, 12/02/01, "2011", by Niall Ferguson, on the
price of oil being out of reach by 2010
New York Times, 2/24/2004, Jeff Gerth, "Forecast of Rising Oil Demands
Challenges Tired Saudi Wells"
Department of Energy Hydrogen Posture Plan, 2/2004, p. A3

Chapter Two: [savefuel.ca] the website of Hydro-Gen, at $199,
experimental patent to make hydrogen by electrolysis of water while motor
is running, Able to legally claim 21% savings on gasoline.
[iags.org] is website for the International Association for Global Security,
who have conferenced with China, and favored Prius +.
[sierraclub.org]—new fuel standards show almost no improvement
[APTA.org] This is the site of the professionals, in the American Public
Transportation Association.

[Calcars.org]—discussions of Prius plus

Chapter Three: *New York Times,* 9/10/05, "The New Prize: Alternative Fuels."

Chapter Four: Transcript from Jim Lehrer News Hour about *The Future of Oil:* , a discussion between Paul Roberts and Daniel Yergin, with Jeffrey Brown interviewing. Inserted were comments by me, that included touching upon the Neele Banerjee report, in the *Times,* "As Oil Prices Soar, OPEC Says 'Not Our Fault.'," June 6, 2004.
A second PBS transcript, from Wall Street Week, "How to Kick the Oil Habit.", by Nicholas Varchaver, of *Fortune* magazine, June 6, 2004.

Chapter Five: *New York Times,* 9/10/05, "The New Prize: Alternative Fuels."
"Ethanols Potential: Looking beyond Corn", by Danielle Murray, in Eco-economy Updates, June 29, 2005.
Natural Resources Research, Dr. David Pimentel and Dr. Tad Patzek, respectively, professors of Ecology and Agricultural Sciences, and of Civil Engineering and Environmental Engineering, Dr. Pimentel at Cornell University in Ithaca, N.Y. and Dr. Patzel at Univ. of California, at Berkeley. The article was in Vol. 15, No. 1, March 2005. " Ethanol Production Using Corn, Switchgrass, and Wood: Biodiesel Production Using Soybean and Sunflower." [pp.65-75]

Chapter Six: None

Chapter Seven: *USA Today,* had an article on 9/9/05, criticizing the people who were writing about hurricanes, as though it was certain global warming result. Ross Gelspan was allowed to rebut that article, where they did not deny global warming, but said we should not be quick to jump to that conclusion. Mr. Gelbspan's article was entitled "Our Denial is at Category 5."
"Evangelical Leaders Swing Influence Behind Effort To Combat Global Warming," *New York Times,* March. 10, 2005
Excerpt from *Christianity Today,* "Heat Stroke," editorial, October, 2004.
Science, "Dilution of the Northern North Atlantic Ocean in Recent Decades." By Ruth Curry, Woods Hole Oceanographic Institution, and Cecile Mauritzen of Norwegian Meteorological Institute, on June 17, 2005.
notjustanotherclone@yahoo.com

Chapter Eight: *International Journal of Hydrogen Energy,* April 2005, "Solar Hydrogen: Environmentally Safe for the Future." By J. Nowotny, C. C. Sorell, L.R. Shepperd, and T. Bak, from the Centre for Material Research in Energy Conversion, School of Material Science and Engineering

University of New South Wales, Sydney, Australia.
International Journal of Hydrogen Energy, April 2005, "Hydrogen as a Fuel: a critical technology?" by S.Z. Baykara, Chemical Engineering Department, Yildiz Technical University.
[Nanosolar.com] has announced their microscopic printing of plastic crystals to improve over PhotoVoltaic Cells. I heard of another company working with <u>plastics</u> who find their material doesn't hold up long, but it was not this company. Dr.David Pimentel's *BioScience* article, 2001, December, "Renewable Energy: Current and Potential Issues", available at *dp18@cornell.edu*
Heliovolt.com
NREL: Newsroom, October 19, 2005

<u>Chapter Nine:</u> *International Journal for Hydrogen Energy*
A lot of my citations are from Harry Braun's book, *The Phoenix Project: Shifting from Oil to Hydrogen—see books.*

<u>Chapter Ten:</u> BMW Group Media Information, 9/2005
Email from the Germany office, from Mr. Kammerer, through Dave Buchko, in BMWNA, in answer to my questions. Material about the early stages of BMW drawn from Harry Braun's book.

<u>Chapter Eleven:</u> Background reference on coal's possible future, "Old King Coal is Back", *Fortune, 2/21/05,* Jeremy Main

<u>Concluding Remarks:</u> H_2Fleet, [phoenixprojectfoundation.us]

ACKNOWLEDGEMENTS

ACKNOWLEDGEMENTS

I wish to express my thanks to the large number of people who have helped me complete this book, in the last ten weeks. People from NASA, BMW, Amtrak, Greyhound, Shell Oil in the disaster period, Detroit Diesel Motor, MTA for Long Island RR, from American Public Transportation Association, Con Edison, Savefuel.com, Calcars.org, American Wind Energy Association, Beacon Press, Woods Hole Oceanographic Institution, Engineers for the environment at several universities, journalists, Institute for Analysis of Global Security, the fine research scientist at a fuel-cell corporation, Dept. of Energy, and others have shown great courtesy and timeliness in getting information for me that makes this book have been an experience that shows the level of professionalism in all these companies.

There are several others I wish to acknowledge: David N. who discussed technical issues, helped me with resource material, and was available to make our efforts stronger, with the two of us working in a considerable degree of unison. The other David who surely seeks solutions to our fuel and global warming problems, and kept up a steady discussion, and was a resource of a great deal of information.

To Professor Dr. T. Nejat Veziroglu. who as president, welcomed me to membership in the International Association for Hydrogen Energy, encouraging any contributions that I might make to help people understand the value of hydrogen as our standard fuel.

To my close friends, and you know who you are, that have heard me develop my learning in this area of energy, respecting what I am trying to do. And make my life a rich one, for their companionship. I especially thank the church that I attend for their great devotion and strength of spirit that comes only from God himself. Your prayers have made the way so much clearer.

To the busy professional men who are trying to read my book, as soon as possible, in case they have comments they would like to make, before the publication process is complete in about five weeks. I could say more, but this book has to come to a definite conclusion

ABOUT THE AUTHOR

Dr. Mary Ann Segal is a woman of multifaceted abilities and interests. Threading through her careers is a common deep concern about the education and well-being of people. She had a career as a teacher in the Antipoverty Programs, and taught in some public junior high schools in New York City. Her personal touch with children and their mothers led her to become a psychiatrist, where her time could be spent with various issues that family therapy included, never dropping the educational needs of the young people. She also showed special skills with several schizophrenic patients. Throughout, she has maintained her interests as a Christian theologian, from time to time writing on theology, especially science and religion. Her expertise on science and religion began when she wrote her honors thesis as a college student, begun at the age of nineteen.

She is a graduate of Swarthmore College, receiving her B.A. in 1962, in religion and premedicine.

Her M.A. in Science Education was from Teachers College, Columbia University in 1965.

She studied theology, church history, and philosophy of religion at Union Theological Seminary between 1963 and 1965.

After teaching three years, she began her study of medicine at the Albert Einstein College of Medicine, receiving her M.D. in 1971.

Between her specialty training and her staff positions she practiced at several major hospitals in the boroughs of New York.

Almost all of her work was centered in disadvantaged areas of the boroughs. Her sudden realization in 2002, that these mothers and children, here and elsewhere in the nation, could possibly face life threatening cold and hunger, because of an impending fuel crisis, which dawned on her early, sent her into emergency mode. She studied both the fuel factors and hydrogen as an energy source, became a member of the International Association for Hydrogen Energy, and began the freelance writing that was apropos of trying to avert, as much as possible, the mounting crisis of the fuel shortage and the search to stop further global warming by introducing the new frontier, the use of hydrogen, from water electrolysis, as our primary fuel. It would be forever renewable and totally clean, as it turns to water when it is used as a fuel.